MARINERS' MEMORABILIA

BRIEF HISTORIES OF
BRITISH DEEP SEA CARGO AND PASSENGER LINES
AND COMMONWEALTH COMPANIES
ILLUSTRATED WITH EXAMPLES OF THEIR
19TH AND 20TH CENTURY CHINA

*Tureen manufactured for the Royal Exchange Shipping Company/
the Monarch Line, probably by Ashworth Brothers.*

VOLUME FOUR

INTRODUCTION

This book, Volume Four, which mostly deals with British passenger and cargo liner companies and a miscellaneous selection of Australian, New Zealand and Canadian lines, together with three earlier volumes, is an attempt to illustrate examples of china used on board British merchant ships and covers the period from the beginning of the 19th century, through to the end of the 20th century. It also gives brief historical details of the companies themselves and the trades in which they were involved.

Whilst reasonably comprehensive, no work of this nature, could ever attempt to list all patterns used on china by any company nor could it hope to cover all companies, particularly those which only existed for a short period of time in the 19th century. Over the period under review there were thousands of different shipping companies, each using during their lifetimes a variety of patterns. For the most part the pieces of china illustrated are examples from the Laister collection and have been acquired over a period of thirty years or so. A few pieces are from the collections of friends. Without doubt, many different examples are in the safe hands of collectors scattered around the world, particularly in North America, although many of the collectors domiciled on the western side of the Atlantic concentrate on the great passenger ship lines, such as the Cunard Steamship Company, White Star Line and French Line.

Many different examples of china are already contained in private collections; others pieces await "discovery" and might be found in the most unexpected places. Sometimes this may be many miles from the sea and in countries where there is no national or obvious reason for their location ie the companies did not trade to the country in which the china was found. This may partially be attributed to ships being sold for breaking up in countries far from their normal routes. A good example of this is the Italian port of Genoa, where there was a thriving shipbreaking industry at the beginning of the 20th century. Other pieces would have been "acquired" over the years as souvenirs.

The shipping companies described in this book traded to all parts of the world and with a very few notable exceptions are no longer in existence. Some restricted their pattern of trading to coastal waters around the United Kingdom. Other companies traded worldwide and owned a variety of types of vessels, passenger liners, cargo passenger liners (usually only carrying twelve or fewer passengers), cargo only and tankers.

Whilst the name "British" is mentioned in the opening paragraph of the introduction, Volumes One and Two also include shipping companies owned in Australia, Canada and New Zealand. The companies were of great importance to the economies of the countries concerned and to what used to be known as the British Empire. Many of these shipping companies were essential for the development of the countries and continents to which they traded, for example Elder Dempster to West Africa, Pacific Steam Navigation Company to Chile and Peru, British India Steam Navigation Company to the Indian sub-continent, Blue Funnel & Glen Lines to Hong Kong and the Far East, Orient Line to Australia and the New Zealand Shipping Company to New Zealand. These prestigious companies formed a link between Britain and her Empire, starting from the very earliest formative years of the countries. Without the major passenger carrying lines such as Cunard, Allan Line and Canadian Pacific to North America, P&O to India and China, Union-Castle to South Africa, Shaw, Savill & Albion and New Zealand Shipping Company to the Antipodes, there would have been fewer emigrants and far from reliable postal services. Expatriates relied on the cargoes and mail brought by sea as a contact with their home countries.

PREFACE

As a small boy in North Yorkshire in the 1940s, and despite the fact there was no seafaring tradition in my family, I decided at a very early age that I wanted to go to sea as a deck officer in the Merchant Navy.

During the course of my school days I commenced collecting photographs and postcards of merchant ships and also the brochures that were issued by the shipping companies. These advertised the delights of such vessels as the **Queen Mary**, **Queen Elizabeth** and **Caronia** and the beautiful liners of the Union-Castle Mail Steamship Company. My obsession with collecting, therefore, dates from the 1940s.

Having completed my education I went to sea as a cadet with the Ellerman's Wilson Line of Hull on their services to Scandinavia, the Mediterranean and the east coast of the United States and Canada. After five years with the Wilson Line I served for ten years with the Union-Castle Line sailing on both their passenger liners to South and East Africa and their refrigerated cargo liners.

Eventually coming ashore my career progressed through a brief period in wharf management followed by 25 years with one of the leading London firms of cargo surveyors and consultants, Perfect Lambert and Company, where I became a partner. As a surveyor I was concerned with inspecting cargo and investigating damage claims all over the world ranging from Patagonia to North Vietnam and most places in between.

Whilst conducting a survey on a small Greek vessel that had been towed into the French port of Brest, following a very serious fire on board, my examination revealed that the engine room of the vessel and the crew accommodation above had been more or less totally destroyed. The intensity of the fire was such that virtually nothing remained intact other than for a small china egg cup which had miraculously survived, this showing the crest of the original German owners of the vessel.

This I took home as a 'souvenir' and from this small beginning has developed a large collection of shipping company china, not to mention some souvenir items and silver plate and glass pieces. My interest in collecting which had remained dormant whilst I was serving at sea is not only an act of acquisition but also one of nostalgia. Every piece I find to add to my collection brings a visual image of the vessel where the china was probably used and the trade upon which the vessels were employed.

Furthermore, my collection has required considerable research into the histories of the more obscure shipping companies, particularly those that existed in the early to mid part of the 19th century - my particular interest. Apart from the maritime aspect, there is also considerable pleasure to be obtained in identifying the manufacturers of the china, which has led to further research.

Peter Laister   Hartley, Kent   March 2021

ACKNOWLEDGEMENTS

A number of friends and fellow collectors have provided access to their collections and this has given me great pleasure and also increased my knowledge of the subject.

Notably, Laurence and Jennifer Dunn, Gravesend, Kent (both now sadly deceased) and Alberto Bisagno of Genoa, Italy, have been particularly helpful. All photographs, with the exception of a small number (which are individually credited) are of the Laister collection and were taken by my wife, Pamela.

Apart from her talents as a photographer she has given me her full support in both this publication and acquiring pieces for my own collection. Without her, my collection would have been very much the poorer and she has faced many hours of "hunting", frequently in very cold, windy and very wet conditions, always with a cheerful heart and given me great encouragement when I started to weaken. My debt to her is immeasurable.

Copyright © 2021 by Peter Laister. The right of Peter Laister to be identified as author of this work has been asserted by him in accordance with the Copyright, Design and Patent Act 1998.

All rights reserved. No part of this publication may be reproduced, stored in a retrieval system or transmitted in any form or by any means (electronic, digital, mechanical, photocopying recording or otherwise) without prior permission of the publisher.

ISBN : 978-1-902953-99-1

Published by Bernard McCall, 400 Nore Road, Portishead, Bristol, BS20 8EZ, England.
Website : www.coastalshipping.co.uk
Telephone/fax : 01275 846178. Email : bernard@coastalshipping.co.uk.
All distribution enquiries should be addressed to the publisher.

Printed by Short Run Press, Bittern Road, Sowton Industrial Estate, Exeter, EX2 7LW.
Telephone : 01392 211909   Fax : 01392 444134
Website : www.shortrunpress.co.uk

CONTENTS VOLUME FOUR

Page

Chapter One 7

Chapter Two 13

BRITISH DEEP SEA CARGO AND PASSENGER LINES

Barrow Steamship Co	14
Black Ball Line of Australian Packets	15
Bowater Steamship Co	17
British Shipowners Co	19
Bullard King & Co (Natal Line of Steamers)	20
Dalgleish R S Limited	22
Egyptian Mail Steamship Co	23
European and Australian Royal Mail Co	24
Euxine Shipping Co	26
Evan Thomas, Radcliffe & Co	27
Galbraith, Pembroke & Co	29
George Gibson & Co	30
George Milne & Co (Inver Line)	32
Great Western Steamship Co	33
Gulf Transport Line (J H Welshford & Co)	34
Hain Steamship Co	35
Harrison J & C	37
Hamilton, Fraser & Company (Inch Line)	39
Indo-China Steam Navigation Co	40
Indra Line (Thos B Royden & Co)	41
Jamaica Direct Fruit Line/Jamaica Banana Steamship Co	42
Japp & Kirby	44
Johnston Line (William & Edmund Johnston)	45
Khedivial Mail Line	46
Knight Line (Greenshields, Cowie & Co)	48
Lancashire Shipping Company (James Chambers & Co)	49
MacIver Line	50
Mogul Line	52
Morison, John & Co	53
Mossgiel Steamship Co Ltd (J Bruce & Co)	54
New York & South America Line (Chas G Dunn & Co)	55
Nicholson J	57
Nova Scotia, Newfoundland and Bermuda Royal Mail SP Co	58
P Iredale & Porter	59
Palm Line	60
Planet Line	62

Rankin, Gilmour & Co	63
Red Cross Line of Steamers to Northern Ports of Brazil (R Singlehurst & Co)	65
Redcroft SN Co (Lewis Lougher)	67
Rio Grande do Sul Steam Ship Co	68
Royal Exchange Shipping Co (The Monarch Line)	70
Ropner & Co	72
Star Line Limited (James P Corry)	74
South American Saint Line	76
Ulster Steamship Co (G Heyn & Sons Ltd)	78
Village SS Co Ltd (R A Mudie & J H Mudie)	81
Warren Line (White Diamond Steamship Co)	82
West India & Pacific Steamship Co	84

Chapter Three

FORMER COMMONWEALTH COMPANIES

Adelaide Steamship Co	86
Anchor Shipping and Foundry Co, Nelson	89
Australasian Steam Navigation Co	90
Australasian United Steam Navigation Co (AUSN)	92
Australian Oriental Line	94
Burns, Philp & Co	95
Huddart Parker	97
Melbourne Steamship Co, Melbourne	100
Northern Steam Ship Co, Auckland	102
Pickford & Black, Canada	103
Quebec Steamship Co	104

CHAPTER 1

TYPES OF CHINA, SUPPLIERS, MANUFACTURERS, DATING AND IDENTIFICATION

TYPES AND SUPPLY OF CHINA

For the most part shipboard crockery had to be strong, durable and capable of withstanding some abuse, either from stewards or the hazards of the sea. During the 19th century and through into the 20th century, almost all crockery manufactured for shipboard use consisted of ironstone china (heavy off-white earthenware). This was somewhat crude by comparison to bone china which was much whiter and finer and which contained between 30% and 50% of bone ash. Bone china was used by some of the more prestigious passenger lines, particularly for 1st class passengers. The decoration was usually achieved by applying transfers, sometimes this being enhanced by hand painting, in colour, over the transfers.

In general, the china used on board was similar or identical to that used ashore and was, after all, "hotel ware" that happened to be used at sea. Most patterns and designs were not specifically manufactured for a particular company. Whilst the china was overprinted with the names of the shipping company, or with their monograms or house flags, identical crockery was often used by more than one line, with the appropriate crest applied. In the cargo or tramp ships owners and in the lower classes on the passenger liners, 2nd and 3rd class and emigrants, the crockery tended to be plain white/off-white with the name of the company imposed. This type of china was frequently "decorated" with a single coloured band running around the rim of the china, often with a pinstripe running parallel just inside the coloured band, in the same colour. Blue was the favoured colour for such decoration, although the colours red and green were sometimes used. Some companies used either blue or red decoration (an example of this is Houlder Brothers). Another popular embellishment, during the middle part of the 19th century, was an anchor chain running around the border of the plate or rim of jugs, usually in a light green or blue colour. A rope border, interspersed with reef knots in a dark blue colour, was also popular at about the same time.

In the case of more decorative china, intricate designs were sometimes favoured and the firm of Minton produced several different designs, these being used by more than one company. An example of this is their "Key Festoon" pattern, which was used by, amongst others, African Steamship Company, British & African Steam Navigation Company, later Elder Dempster, Charente Steam Ship Company (T & J Harrison) and the Chilean line, CSAV (Compañia Sudamericana de Vapores).

Another example of a Minton design is their "Alton" pattern, which was used by the African Steamship Company, Pacific Steam Navigation Company and Canadian Pacific Railway Company. As in the case of the "Key Festoon" design, the single colour of light or dark blue was preferred for the decoration.

Particular striking china (bottom marked) was produced by Copeland Spode for Canadian Pacific Railway Company and later the Allan Line. This was a 'stock' pattern known as "Heron" and came in two versions, multi-coloured and blue.

Not all china was top marked, although most collectors prefer items which are marked on the front or top, the bigger the crest, or company name, the better. However, many of the large passenger liner companies, which preferred to use somewhat more sophisticated or bone china in 1st class, had their china marked with the company name or crest, on the underside. When the company name was printed on the underside, the manufacturer of the china would charge an extra amount to do so. This was for the copper plate required in the process and for the extra print of the name, which had to be taken off the copper plate. The latter remained the property of the shipping company.

There were a number of ship chandlers, or china and glass suppliers, who specialised in supplying the requirements of the ship owners, such as Stonier & Company in Liverpool. These suppliers would hold a stock of replacement china in their warehouses so that when a vessel arrived in port after a poor passage and had sustained a loss of china due to breakages in heavy weather, the suppliers could arrange to replace the china from stock within a day or so. This was preferable to having to wait for replacements to be manufactured.

Apart from Stonier & Company in Liverpool, several chandlers/suppliers existed in the major ports to deal with the ships' requirements. These included Thos F Bennett & Company, Eills & Company, Reynolds & Son, R Livingston, A B Buxton and D A S Nesbitt & Company (all of Liverpool), Cochran & Fleming, A Sneddon & Son Limited, Wylie & Lochhead & Sons Limited, McDougall & Sons Ltd, and Christie Brothers (all of Glasgow), McSymons & Potter, (Glasgow & Swansea), C E Bevington Ltd, and J De Fries & Sons, both London, and C McD Mann & Company, Hanley.

A number of manufacturers, such as Copeland and Davenport, owned their own retail shops and warehouses in the major ports, such as London, Liverpool or Glasgow. Salesmen from these companies would call at the shipping company's office, or sometimes the shipping company's own buyer visited the manufacturers' premises to establish what china was available from stock.

An example of "stock" china, which was used by more than one shipping company, is a Royal Doulton pattern decorated overall with rosebuds, trimmed with a green pinstripe. China of this design was used by the Union-Castle Line, the Royal Mail Lines and the Bank Line. Other examples are a design decorated with flowers, and a tropical bird, used by both the Royal Mail Lines and the Furness Bermuda Line, and a design with small blue flowers overall, used by the Union-Castle Mail Steamship Company (later china was marked British & Commonwealth, after the merger of the company and Clan Line in 1955) and the British India Steam Navigation Company in the 1950s.

Some shipping companies, such as the Orient Line, had their own china specifically designed for them. This company commissioned Wedgwood and their designers to produce patterns for their own exclusive use. An important example of this is a stunning pattern of china decorated in purple and black designed by Edward Bawden. This china, with the pattern name "Heartsease", was introduced with the liner **ORONSAY**, where it was used in the Silver Grill. Another important pattern was designed by Robert Godden. This china with a fine earthenware body was known as "Fouled Anchor" and produced by Wedgwood in their Queensware range, for the P&O-Orient Line in the late 1950s. This was introduced with the liner **ORIANA**.

A particularly striking pattern in brown was designed by Edward Bawden and supplied by Wedgwood, exclusively for the New Zealand Shipping Company. This was used in the 1st class dining saloons.

Lady Casson designed a new pattern of china for the P&O liner **CANBERRA** in 1961. This was manufactured by Messrs Geo L Ashworth & Brothers and the design consisted of a motif of an endless and intricate maze in gun-metal grey.

If a shipping company had a pattern that was exclusive to its own use, it was not uncommon for it to be manufactured by more than one potter. An example of this was china, decorated with seashells in yellow and brown, which was manufactured for the Orient Line in the 1950s and 1960s by Ashworth Brothers, Royal Doulton and the North Staffordshire Pottery Company (Vitrock china).

Whilst most patterns of china were rather plain, understandably so in the case of cargo ship companies and in the 3rd class and emigrant classes of the passenger liners (the greater the decoration, the greater the cost), some companies in the early part and mid-part of the 19th century used very beautiful and ornate crockery. In some instances, the rims of the plates were decorated with flowers and, in the centre, with the portrait of a paddle steamer. Fine examples of these very early pieces are known from the British & American Steam Navigation Company, the Transatlantic Steamship Company, the Eastern Steam Navigation Company (paddle steamer **PRECURSOR**) and several coastal lines, which operated services between Scotland and London. The latter included the London & Edinburgh Steam Packet Company, London and Edinburgh Shipping Company, Dundee, Perth & London Shipping Company and Dundee & Hull Steam Packet Company.

The Pacific Steam Navigation Company commissioned crockery that was particularly colourful. This was decorated in the centre, in colour, with baskets of fruit and the plates had light blue rims. The china was top marked with the company name, in either English or Spanish, the latter being used on the company vessels employed on the West Coast of South America coastal services, between Panama and Valparaiso in Chile. The Union Steam Ship Company of New Zealand was another company using very colourful china, decorated with flowers, this crockery being bottom marked.

MANUFACTURERS

The leading potters/manufacturers of the time were employed to produce the china. Some are still household names in the United Kingdom, although recently a number have shifted their production to the Far East or eastern Europe where the cost of labour is much cheaper.

Examples of 19th century manufacturers who produced shipping company china are by:-

John & George Alcock, Corbridge	1839-1846
G L Ashworth & Brothers, Hanley	1862-1968 (Mason's Ironstone from 1968)
J & M P Bell & Co Ltd, Glasgow	1842-1928
Edward F Bodley & Co (or Son), Burslem	1862-1898
Booths Limited, Tunstall	1891-1948 (later Booths & Colclough Ltd)
Sampson Bridgwood & Son, Longton	1805 -
Brown-Westhead, Moore & Co, Hanley	1862-1904
Clementson Brothers, Hanley	1865-1916
Copeland & Garrett, Stoke	1833- 1847
Copeland W T & Sons Ltd, Stoke	1847 (now Copeland Spode)
Davenport, Longport	1793-1887
Doulton & Co, Lambeth etc	1858-1956
Dunn Bennett & Co, Burslem	1875 -
James Edwards & Son, Burslem	1839-1841
Thomas Hughes, Longport, Burslem	1895-1957
Johnson Bros (Hanley) Ltd	1899-
Kerr & Binns, Worcester	1852-1862
M King, The Pottery, North Shields	
John Maddock & Sons Ltd, Burslem	1855 -
C T Maling & Sons Ltd, Newcastle upon Tyne	1890-1963
Mann & Company, Hanley	1858-1860
Minton, Stoke on Trent	1793 -
Francis Morley & Co, Shelton, Hanley	1845-1858
John Ridgway & Co, Shelton, Hanley	1830-1855
Ridgwood, Morley, Wear & Co, Shelton, Hanley	1836-1842
Wedgwood & Co Ltd, Tustall	1860-1965
James F Wileman, Longton	1869-1892
Arthur J Wilkinson Ltd, Burslem	1885-1947
F Winkle & Co Ltd, Stoke on Trent	1890-1931

In addition to some of the potters listed above, who produced china in both centuries, the following companies manufactured shipping company china in the 20th century:-

William Adams & Sons (Potters) Ltd, Tunstall & Stoke	1769-
Adderleys Ltd, Longton	1906 -
Chas Allertons & Sons, Longton	1859-1942
John Aynsley & Sons Ltd, Longton	1875 -
Bakewell Brother Ltd, Fenton	1927-1943
E Brain & Co (Foley China), Fenton	1903-1963
British Anchor Pottery Co Ltd, Longton	1884-
Cauldon Limited, Shelton, Hanley	1905-1920
George Clews & Co Ltd, Tunstall	1906-1961
Dudson Bros Ltd, Hanley	1898-

Furnivals Ltd, Corbridge	1890-1968
Gibson & Sons Ltd, Burslem	1885-1985
Grindley Hotel Ware Co Ltd, Tunstall	1908-
Hammersley & Co	1887-
Jackson & Gosling Ltd, Longton	1886-1961
A B Jones & Sons Ltd, Longton (Royal Grafton)	1900-1972
Keeling & Co (Losol Ware)	1886-1936
J & G Meakin Ltd, Hanley	1851-
Myott, Son & Co Ltd, Corbridge & Hanley	1898-
North Staffordshire Pottery, Corbridge & Hanley (Vitrock)	1940-1952
R H & S L Plant Ltd, Longton (Tuscan China)	1898-
Pountney & Co Ltd, Bristol	1849-1969
Ridgway Potteries Ltd, Stoke on Trent	1955-1964
James Sadler and Son Ltd, Burslem	1899-
Royal Doulton formerly Doulton, Lambeth	1882-
Josiah Wedgwood & Sons Ltd, Burslem	1759-
H M Williamson & Sons, Longton	1879-1941

These potters were for the most part situated in the Stoke on Trent area of the English Midlands, "The Potteries".

DATING AND IDENTIFICATION

Dating and identifying china used on board ships can be very frustrating and covers the full spectrum, from easy to difficult, if not impossible. Most collectors have a box hidden away, where pieces accumulate dust, pending identification. Some of these may turn out to be treasures, most are likely to be of no interest.

Of course, the easiest pieces to identify are those bearing the name of the company, or a company house flag, and from a collectible point of view these are the most desirable. However, even if the pieces are marked in this way, it can still be difficult to learn much about the company when the china dates from the 19th century. This is especially the case if the company only survived for a year or so.

Books of House Flags

Books detailing shipping company flags and liveries are essential tools. Of these, three books published by Lloyds of London are the most invaluable when dealing with early pieces. These books, *Lloyd's List of House Flags* published in 1882, and *Lloyd's Book of House Flags & Funnels* published in 1904 and 1912, are themselves very collectible. Contemporary to these, are books published by Griffin & Company, Portsmouth - *Flags National & Mercantile House Flags & Funnels*, and by Thomas Reed & Co Ltd, Sunderland. In the case of the latter publisher, their *S. S. House Flags* and *Reed's Flags and Funnels* are of particular use, when dealing with the more obscure cargo ship and tramp companies of North East England.

Other more recent books are those published by Brown, Son & Ferguson, Ltd, Glasgow, *Brown's Flags & Funnels*, which now runs into nine editions, and *Flags, Funnels and Hull Colours* published by Adlard Coles Limited.

The Liverpool Journal of Commerce published a number of charts depicting House Flags and Funnels, over a period of several years. Two other noteworthy books on the subject are *House Flags and Funnels of British and Foreign Shipping Companies*, drawn and edited by E C Talbot-Booth in 1937 and *A Survey of Mercantile Houseflags & Funnels* by J L Loughran, published by Waine Research Publications in 1979. This book also gives interesting details concerning the shipping companies, and how their house flags evolved. All the above are helpful, but unfortunately there are many companies which are not recorded, this being particularly true of the early 19th century lines and the less prestigious companies.

DATE AND REGISTRATION MARKS

Date and registration marks on the china are also invaluable, but rarely appear. White Star Line china in the first decade of the 20th century was sometimes marked on the underside with the month and year, as was Cunard Steamship Company ivory ware china manufactured by John Maddock & Son. Other more recent pieces carry an impressed date mark.

A diamond shaped mark was impressed or printed on Victorian china between 1842 and 1883. These marks recorded the class of china, the day of the month, the month, the year and parcel number. From 1884 registered designs were numbered consecutively, with the prefix "Rd. No (number)". Again, this information is invaluable, with the proviso that it is only one part of the equation. For example, the dates and numbers show when the design or pattern was first registered. However, some patterns, like Minton's "Key Festoon", were used over several decades and could give a false impression as to when the china was in use by the shipping company. That is, the design was registered before the shipping company came into existence.

Apart from the above, many of the leading manufacturers had their own system of date marks and these can also be of great assistance.

Undoubtedly the most invaluable source of information when it comes to china marks including dates is the "bible" of collectors, the *Encyclopedia of British Pottery and Porcelain Marks* by Geoffrey A Godden, published by Barrie & Jenkins. This details and illustrates many of the marks used by china manufacturers both large and small. It is very comprehensive but even this great work cannot illustrate all the marks and variations which occur.

An indication as to date of manufacture can also be obtained when the marks give the name "England" or "Made in England", The former was added to marks from 1891 and the latter indicates a 20th century date. This does not always hold true, as Ashworth Brothers continued to show the mark "England" into the 20th century, and not the latter mark.

MONOGRAMS

Without doubt, china that is top marked with a monogram or intertwined letters can be the most difficult to identify, unless the monogram is surrounded by the company title, usually in a garter. An example of this is the "CNCo" monogram used on early china by the China Navigation Company. Given the absence of the shipping company name, or a house flag, on the piece, one has to resort to deduction which can be all too fallible. In this situation, company literature or letterheads can be the answer to the collector's dilemma, as frequently the same monogram or logo is used on both the tableware and written material such as letterheads or brochures.

<u>Named vessels</u>

For the most part, ship's tableware was not marked with the name of a particular vessel; it just did not make economic sense to do so. Retaining a stock of "named" china in a warehouse was a costly business, particularly for companies whose vessels carried a large number of passengers. Like all generalisations, however, there were exceptions. In the 19th century a number of companies had very ornate china manufactured for them. This bore the name of the ship, and sometimes a portrait of the supposed vessel, usually a paddle steamer. These "portraits" were, to say the least, often inaccurate and frequently contained a large amount of artistic licence.

As the years elapsed the shipping companies became controlled more and more by accountants, rather than by owners who were aware that shipping had always had good and bad years and accepted this. In an effort to reduce costs many stopped using marked china and relied on "off the shelf" purchases of stock patterns. There were honourable exceptions to this and most of the remaining prestigious passenger liner and cruising companies still have china either designed for them or top or bottom marked with their name or logo.

In the case of other vessels, china was specially manufactured for the ship when she was built. An example of this, is a pattern which resembled gold "snowflakes". This was used by the Cunard Line on the **QUEEN ELIZABETH 2**, when she was brought into service. In some companies, special china was commissioned for specific dining areas on the vessels, such as grill rooms. Here the passengers were charged a supplement for the pleasure of dining in more elegant surroundings, or receiving extra special care and attention. An example of this was very striking china with a rich blue border which was manufactured by Copeland Spode for the grill rooms of the White Star Line vessels **OLYMPIC** and **TITANIC**. Pieces of this china are very hard to find and command very high prices when they come on to the market. Other examples of exclusive china were to be found in the Queen's Grill of the **QUEEN ELIZABETH 2** and the grill rooms of the Orient Steam Navigation Company's liners.

China printed with the name of specific vessels was not confined to the passenger liner companies. Many of the shipping companies, such as the owners of cargo liners, tramp ships and colliers provided their vessels with china which bore the name of the vessel. Almost invariably this china was plain, being decorated at the rim with one or more coloured bands (usually blue, sometimes red) and the name of the vessel. This china was sometimes marked with a house flag within a garter or similar surround.

Some indication as to where and when tableware was used and on which ships, in what class and period of time, may be established by studying official postcards and brochures. These were issued by the shipping companies to illustrate the dining areas on their vessels. Regrettably some of these illustrations, whilst depicting the dining saloons, are not clear or sharp enough to determine the patterns on the china, if there is any china in the picture. Nevertheless, it is always worth checking these illustrations.

Unfortunately, china patterns are sometimes attributed to a specific class, or vessel, when the provenance of this attribution, which may be based on false premises or speculation, cannot be relied upon. This is particularly so in the case of earlier pieces of china dating from the 19th century. With modern vessels the china patterns are well documented and company literature and other pieces of ephemera and books are profusely and clearly illustrated with photographs showing the dining areas on board and the table settings.

PRICES

No prices are given in this work as to some extent prices are meaningless and can fluctuate both up and down. This is because prices are dependent on a number of variable factors, such as supply and demand and whether or not companies are "fashionable" at the time. Clearly pieces from the 19th century are harder to come by and understandably command higher prices.

Recently, online auctions and in particular eBay, have changed the market considerably to the advantage and also disadvantage of the serious collector. The internet has opened up the world to collectors, so that later and more common pieces are easier to come by. This has to some extent depressed the prices of such pieces. Even the prices of White Star Line china have come down somewhat, from the very high prices and the buying frenzy that prevailed after the film "Titanic" was released. On the other hand, serious collectors now have the opportunity of acquiring rare and older pieces online, although the competition for these pieces is usually fierce.

CHAPTER TWO

BRITISH DEEP SEA CARGO AND PASSENGER LINES

Barrow Steamship Co
Black Ball Line of Australian Packets
Bowater Steamship Co
British Shipowners Co
Bullard King & Co (Natal Line of Steamers)
Dalgleish R S Limited
Egyptian Mail Steamship Co
European and Australian Royal Mail Co
Euxine Shipping Co
Evan Thomas, Radcliffe & Co
Galbraith, Pembroke & Co
George Gibson & Co
George Milne & Co (Inver Line)
Great Western Steamship Co
Gulf Transport Line (J H Welshford & Co)
Hain Steamship Co
Hamilton, Fraser & Company (Inch Line)
Harrison J & C
Indo-China Steam Navigation Co
Indra Line (Thos B Royden & Co)
Iredale & Porter
Jamaica Direct Fruit Line/Jamaica Banana Steamship Co
Japp & Kirby
Johnston Line (William & Edmund Johnston)
Khedivial Mail Line
Knight Line (Greenshields, Cowie & Co)
Lancashire Shipping Company (James Chambers & Co)
MacIver Line
Mogul Line
Morison John & Co
Mossgiel Steamship Co Ltd (J Bruce & Co)
New York & South America Line (Chas G Dunn & Co)
Nicholson J
Nova Scotia, Newfoundland and Bermuda Royal Mail SP Co
Palm Line
Planet Line
Rankin, Gilmour & Co
Red Cross Line of Steamers to Northern Ports of Brazil (R Singlehurst & Co)
Redcroft SN Co (Lewis Lougher)
Rio Grande do Sul Steam Ship Co
Royal Exchange Shipping Co (The Monarch Line)
Ropner & Co
Star Line Limited (James P Corry)
South American Saint Line
Tomlinson & Thompson
Ulster Steamship Co (G Heyn & Sons Ltd)
Village SS Co Ltd (R A Mudie & J H Mudie)
Warren Line (White Diamond Steamship Co)
West India & Pacific Steamship Co

BARROW STEAMSHIP COMPANY LTD

The Duke of Devonshire owned much of the property in Barrow-in-Furness, including a steel company (which made large quantities of railway lines destined for North America) and the Barrow Shipbuilding Company. In addition, he was involved with developing the dock system at Barrow. He had discussions with the Henderson Brothers, owners of the Anchor Line, regarding the possibility of starting a steamship Line between Barrow and New York and the Barrow Steamship Company was created in 1872 for this purpose. The Duke of Devonshire subscribed £100,000 of capital to this venture and the Henderson family a similar sum, management of the company being entrusted to the Anchor Line. During the lifetime of the company the ships of both companies were interchanged between the two fleets, as and when required.

Company crest and motto.

In 1874 the Anchor Line vessels **ETHIOPIA** and **BOLIVIA** were sold to the Barrow Steamship Company, although they were still employed on Anchor Line services. The first ship built for the line was the **ANCHORIA** (4,168grt) built by the Barrow Shipbuilding Company in 1875 and initially she sailed for the Anchor Line on their Glasgow to New York service. The inaugural sailing from Barrow to New York was made by the **CASTALIA** in 1880.

The **CITY OF ROME** was a notable iron vessel built by the Barrow Shipbuilding Company for the Inman Line, however, it transpired that she was underpowered for the fast transatlantic service. After five voyages between Liverpool and New York she was rejected by the Inman Line in May 1891 and handed back to her builders, ownership then being transferred to the Barrow Steamship Company and management to the Anchor Line operating on their route from Glasgow to New York.

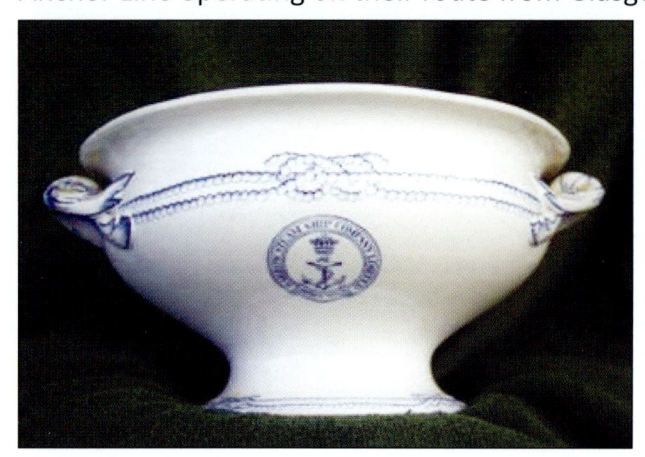

A large soup tureen or/fruit bowl, manufactured by Brown Westhead Moore & Co. Circa 1880. The rope and knot was a popular decoration on shipping company china in the 19th century.

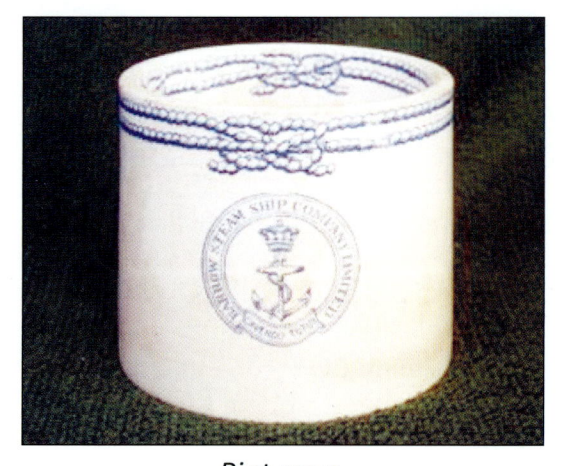

Pint mug.

The **CITY OF ROME** completed her last transatlantic voyage in 1901, by which time all other Barrow Steamship Company ships had been transferred to the Anchor Line or scrapped and the same year the Barrow Steamship Company Limited was wound up.

BLACK BALL CLIPPER LINE OF PACKETS

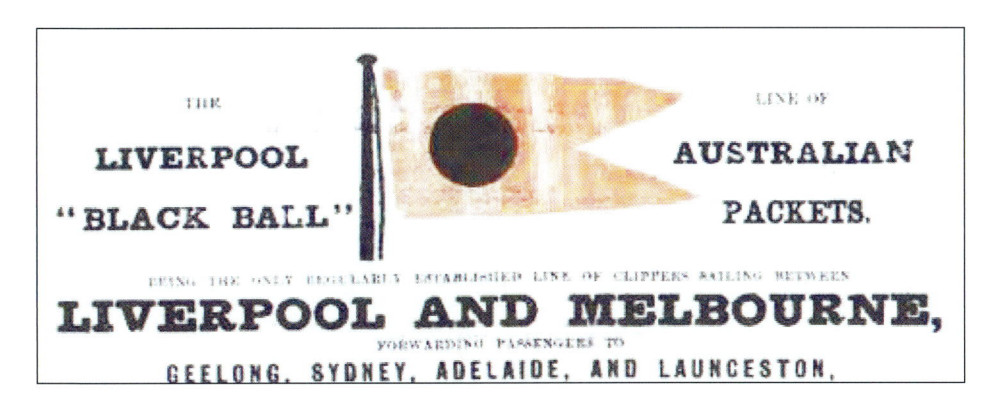

Without doubt the Black Ball Line of Clipper Packets was one of, if not the most famous, sailing ship companies of the 19th century. It was closely connected to the growth of emigration to Australia.

It was founded in Liverpool in 1852 by James Baines and Thomas MacKay with two other partners, Joseph Greaves and John Taylor, to run a regular packet service to Australia. The discovery of gold during the latter part of 1851 and the frenzy that resulted, led to a demand for passages to Australia and fuelled emigration and the need of a regular and reliable service was obviously evident to the partners. Obtaining a mail contract probably added to the incentive to provide larger and faster vessels for the service to Australia.

Crest of the Black Ball Line Packets as depicted on their china and letterheads.

In 1852 the company acquired the first of a number of fast wooden Canadian and American built clipper ships - the **MARCO POLO** (1851/1,625 tons). She was built by James Smith, St John, New Brunswick, Canada. Amongst other notable and well-found wooden ships purchased by the partners were several built by Donald McKay, East Boston, USA:-

> **CHAMPION OF THE SEAS** (1854/2,447 tons)
> **LIGHTNING** (1853/2,083 tons)
> **JAMES BAINES** (1854/2,515 tons)
> **DONALD McKAY** (1855/2,604 tons)

Apart from wooden ships, the firm also owned a number of vessels built of iron and the auxiliary steamer **GREAT VICTORIA** (2,320 tons). She had been built in Nantes, France, in 1854 and previously named **JACQUARD** and was bought by the Black Ball Line in 1863.

Her maiden voyage for the line was from Liverpool to Melbourne the same year. Although the vessels were sailed by competent and experienced masters and crews, the voyage could be somewhat hazardous and a number of ships were lost due to heavy weather, fires and collisions with other vessels or icebergs. On safe arrival at their destination many of the Commanders of the vessels were awarded tokens of esteem by their grateful passengers. This was clearly so on the maiden voyage of the **GREAT VICTORIA**, when her master, Captain James Price, was presented with silver plated tea and coffee pots that were engraved with the following inscription:-

*"Presented to Capt James Price by the Saloon Passengers of the **S S GREAT VICTORIA**
on her First Voyage from Liverpool to Melbourne in token of regard and esteem 1863".*

*Electroplated tea and coffee pots manufactured by Elkington and the engraving on them,
detailing the presentation to Captain James Price.*

Ownership of most of the Black Ball Line fleet of ships was split between James Baines & Company and T M Mackay & Co, the rest of the ships being chartered from other owners as they were needed. These vessels were well equipped with staterooms for cabin passengers, and saloons, smoking rooms and well ventilated accommodation for steerage passengers and emigrants. Many were record breakers on the long route between the United Kingdom and the Antipodes.

For the most part the fleet traded to Australia and sometimes to New Zealand. Other voyages took the vessels to China and on occasion, some were employed on Government service as troop transports to places as diverse as the Crimea. However, by 1865 emigration had declined and the company considered the possibility of developing their interests in steamships. To this end in 1864 it was proposed to start a steamship line to Australia, the Australian & Eastern Steam Navigation Company. This service would include the Black Ball Line, White Star Line and Gibbs, Bright & Company. Unfortunately, this venture proved to be a commercial disaster. Mortgages were sought to keep the company operating, some of these being obtained from Barned's Bank and when this bank collapsed the Black Ball Line was forced into liquidation. The company liquidators sold off 40 of the 61 ships owned by the company in 1866. These sales included all the American built clippers and the auxiliary **GREAT VICTORIA** although the **MARCO POLO** was retained.

The Black Ball Line managed to keep trading until 1871; however, the great days of the line were over.

A fine painting of **GREAT VICTORIA**

BOWATER STEAMSHIP COMPANY LIMITED

The Bowater Steamship Company Limited was founded in 1955, the brainchild of Sir Eric Bowater. However, the company's interest in shipping commenced in 1938 when Bowaters, a leading manufacturer of pulp and newsprint, acquired the Corner Brook, Newfoundland, operation of the International Power & Paper Co of Newfoundland. This became Bowater's Newfoundland Pulp & Paper Mills Ltd. The deal included two deep sea vessels, **HUMBER ARM** and **CORNER BROOK**, which ran between the mill, east coast of USA ports and the United Kingdom. These ships were managed on behalf of Bowater by Furness Withy.

Newfoundland - Corner Brook Pulp & Paper Mill.

Prior to this the company had opened a paper mill in 1926 on the Thames at Northfleet and another four years later at Ellesmere Port in Cheshire and in 1936 obtained an interest in a mill at Kemsley and Sittingbourne, Kent. This was served with pulp imported through the small nearby port of Ridham Dock.

Dinner plate made by J Maddock, supplied by Stonier & Co, Liverpool.

A later acquisition in 1956 (although Bowater had first tried to buy the company in 1936) was the Mersey Paper Company Limited of Liverpool, Nova Scotia. This company also owned ships which were managed by the Markland Shipping Company, examples being the **MARKLAND** and **LIVERPOOL PACKET**, which in due course were transferred to the Bowater Steamship Co. The **HUMBER ARM** was torpedoed in 1940 and a number of war replacement vessels were also sunk, the fleet suffering grievous losses.

Following the war, and by the time of the formation of the Bowater Steamship Company in 1955, the company calculated that it needed to ship a million tons of wood pulp/raw materials from mills in Scandinavia and Canada every year. The decision was taken, therefore, that three medium sized and six smaller ships should be ordered. The first of the medium sized ships was the turbine steamer **MARGARET BOWATER** delivered in 1955 and her sister **SARAH BOWATER** (both 6,500grt approx). The first of the smaller motor ships was the **ELIZABETH BOWATER** (4,000grt approx), followed by the sisters **CONSTANCE**, **ALICE**, **GLADYS**, **PHYLLIS** and **NINA BOWATER**.

Egg cup made by Ashworth Brothers and silver plate menu holder.

These vessels were managed from 1957 by Cayzer, Irvine & Co Ltd - later the British & Commonwealth Shipping Company and employed providing the various mills with raw materials, such as wood pulp from Scandinavia. In addition, regular shipments of newsprint were made to Alexandra, Washington, USA from the Mersey Mill, Liverpool, Nova Scotia for the 'Washington Post' and from Corner Brook, Newfoundland via the Great Lakes to Chicago with newsprint for the 'Chicago Tribune'.

The Bowater Steamship Company was, sadly, short lived as the parent company went into decline and by 1977 all company owned ships had been disposed of, the mills being served by cheaper chartered tonnage, as required, at prevailing market rates.

*A fine painting of **ALICE BOWATER** by Liverpool artist Philip Welsh.*
Of 4045grt she was completed in 1959 by Cammell Laird at Birkenhead.

The last vessel to be disposed of was the **NINA BOWATER**. These ships were run like miniature liners, with 'silver service' in the dining saloon.

BRITISH SHIPOWNERS CO LTD

The British Shipowners Company was founded in Liverpool by James Beazley in March 1864 and during its lifetime it owned a number of notable iron sailing and steamships. These vessels were primarily employed on the Far Eastern, Indian and Australian trades out of London and Liverpool and were built specifically for time charter to well-known British companies. These included Anchor Line, American Line, New Zealand Shipping Company, Shaw Savill & Albion and Furness Lines.

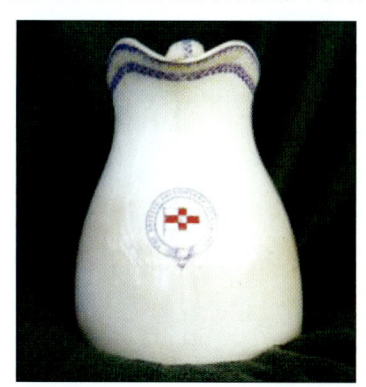

Large water jug with blue purple decoration showing the company house flag.

Prior to the formation of the British Shipowners Company, James Beazley had already had a successful career as a private shipowner and operated iron sailing ships both in the Australian emigrant trade and in the China tea trade. Two of his vessels were sold to the new company in 1864 - the **NELSON** and the **GLENNA** which was almost immediately lost whilst on passage to China on her maiden voyage in 1864.

The first new vessels owned by the company were two iron ship-rigged sailing ships built in 1864 - the **BRITISH INDIA** (1,199nrt) and **BRITISH SOVEREIGN** (1,345nrt). This system of naming set the pattern for all future company vessels and more or less all subsequent ship names began with the prefix BRITISH. By the late 1870s the company owned one of the largest British fleets of sailing ships, the vessels increasing in size year by year.

The first steamships were introduced in 1878/9, the **BRITISH EMPIRE** (2,153nrt/1878) and the **BRITISH CROWN** (2,245nrt/1879). After this the sailing ships were gradually phased out and by 1894 the fleet consisted of ten vessels, four of which were sailing vessels. Whilst the Far East and Australia had seen many of the company vessels in the early years, from the late 1870s much of the fleet was time chartered to companies operating on the North Atlantic, notably to the American Line. Several vessels were built to serve as liners for this trade and when disposed of these were bought by well-known liner companies such as Holland America Line.

Dinner plate with blue decoration at rim T/M with coloured house flag. Thos Bennett, Liverpool.

From July 1895 the company was managed by Gracie, Beazley & Co until its demise in 1906. The last steamship to be owned by the company was the **BRITISH TRADER**. She was sold in 1906 to Russia and renamed **SOTRUDNIK**. Later in 1908 she passed to the Russian East Asiatic Co, St Petersburg, and was renamed **RUBONIA**. She was sunk in 1915 by the German submarine U.36 on voyage Cardiff to Archangel.

BULLARD KING & CO
(THE NATAL LINE OF STEAMERS)

Bullard King & Co was formed in 1850 by two sailing ship masters, Samuel Bullard and David King, to operate sailing vessels in the Mediterranean trade. In 1860 direct services were commenced from London to Port Natal (Durban) in South Africa around the Cape of Good Hope under the name The Natal Direct Line. The company's first steamer was the **PONGOLA** of 1879, followed by several other vessels. In 1883 four new steamers were introduced - these being given Zulu names with the prefix "UM" eg **UMTATA**, a system of naming which continued until the demise of the company in 1960. From 1889 to 1911 the company had regular sailings between South Africa and India, mainly carrying Indian labour for the South African sugar plantations.

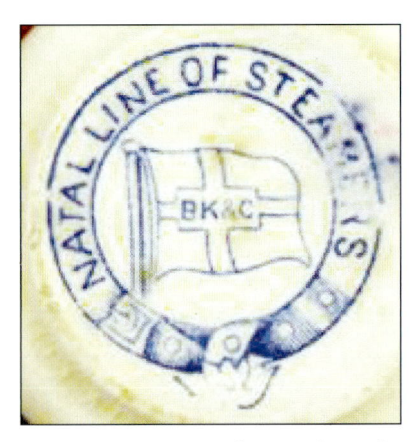

Coffee cup showing the Natal Line of Steamers crest on the underside.

In May 1919 Bullard King was taken over by the Union-Castle Mail Steamship Company, although the Natal Line retained its identity and livery of a grey hull with a buff, black topped funnel and chocolate band. During the ensuing years several vessels were interchanged with the parent company, a practice that continued after 1955 when the Union-Castle line merged with the Clan Line and became the British & Commonwealth Shipping Company. Any vessels that were transferred from Union-Castle, and later the Clan Line, were given traditional "UM" names.

Cover of a company brochure.

A cereal bowl manufactured by Masons.
The pattern is "Fruit basket", a stock pattern of the manufacturers circa 1950s.

The larger vessels such as **UMTATA** and **UMZINTO** carried approximately 100 passengers and provided comfortable accommodation for them. These passengers were looked after by Indian crews and were a popular alternative to sailing on the larger mail ships, operated by the parent company.

By 1959 the Natal Line was on the edge of closure. However, it then evolved into the Springbok Line Ltd (a subsidiary company - the Springbok Shipping Company had been set up by Union-Castle some ten years previously) and in 1960 the company was brought into life using the six remaining Bullard & King ships. These were later given Springbok names with the suffix of "BOK" eg **GEMSBOK**. After eighteen months the line was then absorbed by the South African Marine Corporation (SAFMARINE).

An early Natal Line steamer.

21

R S DALGLEISH LTD

Dalgleish was one of the many north east coast based tramp ship owners, although the company vessels were not all employed on deep sea trades as tramps and some were employed as colliers on the coastal and the near continent trades.

Robert Stanley Dalgleish formed his company in 1906, when he bought an eleven year old ship from a London owner. This ship was renamed **KENILWORTH** and set the pattern of naming for all future vessels, with the suffix of WORTH at the end of their names. The company entered the First World War with three tramps, the **KENILWORTH**, **HAWORTH** and **WENTWORTH**. All were to become war casualties, one being mined and two torpedoed. However, several new and second hand vessels were purchased during the war and at the time of the armistice the company fleet consisted of five deep sea tramps and a pair of colliers. A rapid expansion took place after the war, the vessels being owned under several company names (in particular the Watergate Steam Ship Co Ltd). The company operated two separate fleets namely tramps and collier/short sea traders - the latter being operated under the name of 'Robert Stanley Shipping Company Limited'. This company was wound up in 1929 and ownership of their vessels transferred to Dalgleish.

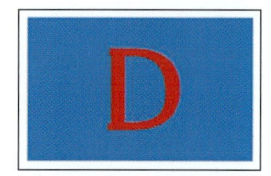

The company houseflag.

The tramp ships were traded worldwide and in the summer of 1932 the company started in the ice free season, what was to become an annual trade (with the exception of the period of the Second World War) to the Canadian Hudson's Bay port of Churchill. In fact, during the years of depression this kept a number of vessels employed and was most beneficial to the company's finances. The first ship to bring grain from Churchill was the **PENNYWORTH**. By 1939 the company was operating four tramps and four colliers and at the end of the Second World War, the fleet stood at one deep sea tramp and three colliers.

In the 1950 and 1960s it became a custom to commission commemorative pieces of china each year showing the month the first company vessel arrived at Churchill, such as the pieces illustrated below. These were made by Grimwades.

Commemorative goblet issued by the company in July 1963 and jam pot dated July 1970 showing the Dalgleish houseflag and Churchill (the port the vessels loaded at in Hudson's Bay).

Amongst the vessels acquired in 1950 was the **FORT DAUPHIN** a Canadian-built war time standard ship of 7,133grt built in 1943. She was renamed **WARKWORTH** (the third ship of this name in the fleet) and for the next seven summers she was employed on the trade to Churchill - winning the ice race on several occasions, when her master was awarded with the customary cane walking stick for the first ship to arrive.

In 1969 the Watergate Steam Ship Co Ltd was sold to the investment company, Lonhro, the fleet of three tramp ships, two ore carriers and a bulk carrier on order at the time all being taken over by Lonhro, although management of the fleet remained with Dalgleish. Kristian Jebsen UK Ltd bought the company from Lonhro in 1974, management being retained by R S Dalgleish. However, by 1979 all vessels had been disposed of and R S Dalgleish went into voluntary liquidation the same year.

EGYPTIAN MAIL STEAMSHIP COMPANY LIMITED

The Egyptian Mail Steamship Company Limited was a short lived British company that operated fast mail and passenger services in the first decade of the 20th century between Marseilles, France and Alexandria, Egypt.

The most celebrated of the company's vessels were the liners **CAIRO** (11,117grt) and **HELIOPOLIS** (11,146grt). Both vessels were built by the Fairfield Shipbuilding and Engineering Company, of Govan. **CAIRO** was launched in 1907 and entered service in January 1908.

1907 Advertisement and share certificate.

Company silver plated muffin dish made by James Dixon.

Company crest.

The service was not successful and both ships were laid up in 1909, when the service ended. The two ships were then sold in 1910 to Canadian Northern Steamship Company, a subsidiary of the Canadian Northern Railway Company. They were then refitted for a service on the North Atlantic as the **ROYAL EDWARD** (ex **CAIRO**) and **ROYAL GEORGE** (ex **HELIOPOLIS**) operating between Montreal/Halifax and Avonmouth. The **ROYAL EDWARD** was sunk during the First World War with a large loss of life whilst transporting Commonwealth troops.

Souvenir spoon from the ROYAL EDWARD.

EUROPEAN & AUSTRALIAN ROYAL MAIL COMPANY

Crest of the European & Australian Royal Mail Company.

This short lived company was founded in 1856 when, following the end of the Crimean War, the Admiralty called for tenders to operate a mail service to Australia. The mails had to be carried between Suez and Melbourne via Point de Galle and the tender was subject to very strong penalties if the service was delayed or not complied with.

Prior to this a company had been formed in Glasgow in 1853 by Robert Henderson & Co for the purpose of using surplus ships to serve a route to the Californian Gold Fields via the isthmus of Panama. This company was the European & Colombian Steam Navigation Company and was to employ two screw steamers on the route, the **EUROPEAN** and the **COLUMBIAN**. However, by the time these vessels were released from Black Sea trooping duties, the Californian Gold Rush was over and the proprietors turned their thoughts to tendering for the new mail contract to Australia, which they succeeded in obtaining. Neither the **EUROPEAN** nor **COLUMBIAN** were found to be suitable vessels for this new service and the first sailing was made in February 1857 by the chartered steamers **SIMLA** and **ONEIDA** (2,293grt/1855). The latter was purchased from the Canada Ocean Steam Ship Company. Other chartered vessels were used, pending the delivery of three new steamers - **AUSTRALASIAN**, **COLUMBIAN** (both built in 1857) and the **TASMANIAN** (1858).

Initially surplus Cunard Line vessels were employed on the Mediterranean sectors between Southampton, Malta and Alexandria and between Marseilles and Malta. By the end of 1857 the company was already in serious financial difficulties and the Mediterranean sectors were sub-chartered to the Royal Mail Steam Packet Company. In 1857 complaints about the conduct of the line were being heard in the House of Commons and the London press. The company was unable to maintain the monthly schedule due to lack of funds and financial penalties and became bankrupt in mid-1858. For a few more months the mail contract to Australia was maintained by the Royal Mail Steam Packet Company, until a new tender was obtained by the P&O Steam Navigation Company in 1858.

Tureen manufactured by J & M P Bell.

Saucer of the tureen.

A silver plated candle holder made for the company by Elkington & Company.

***AUSTRALASIAN** (1857/2,902grt) built by J & G Thomson, Glasgow.*

25

EUXINE SHIPPING CO LTD, LONDON

The Euxine Shipping Company was owned by the Dutch van der Zee family and registered in London in 1932, having previously been registered in the Turkish port of Smyrna (now Izmir). Prior to this the family had successfully operated the Euxine Shipping Company running cruise ships along the Nile in Egypt. They were also merchants and shipowners, their companies including Reederij W H van der Zee, Smyrna, which evolved into the company based in London.

Company ashtray made by Jas Green & Nephew showing the houseflag on one side and a vessel on the reverse.

The vessels owned by the company were quite small and served in the areas of the Mediterranean and Black Sea. Regular 'liner' services were run to the United Kingdom, the ships frequently loading esparto grass for the United Kingdom, which was used in the production of bank note paper. Following the Second World War four similar sized vessels entered the fleet in 1946, 1947 and 1948 - **HENDRIK** (46/2270grt), **HELKA** (47/2111grt), **HENDI** (47/2023grt) and **HENZEE** (48/2372grt). These vessels each had passenger accommodation for twelve passengers.

***HENDRIK** built in 1946, 2,270grt.*

EVAN THOMAS, RADCLIFFE & COMPANY, CARDIFF

In the latter part of the 19th century and well into the 20th century the Welsh city of Cardiff became one of the largest ports in the British Empire and the home port for many tramp and cargo ship owners. The reason for this prosperity was the result of the port's proximity to the abundance of good quality coal in the nearby Welsh coalfields which was exported in large quantities notably to ports in the Mediterranean and South America.

The largest of the locally based shipowners was the firm of Evan Thomas, Radcliffe & Company which was founded in 1881 by a Ceredigion Master Mariner, Evan Thomas, and a Merthyr Tydfil businessman, Henry Radcliffe, partially with their own capital and partially with money raised locally in Wales. The first vessel to be owned by the company was the schooner rigged iron steamer **GWENLLIAN THOMAS** (1882/1082grt) built by Palmers Shipbuilding and Iron Company of Jarrow which was delivered to Cardiff on 24 June 1882 and sailed with a cargo of coal for St Nazaire, returning to Cardiff with iron ore from Bilbao. She was commanded by Capt Evan Thomas. This vessel was named after Captain Thomas's daughter.

*A typical early Evan Thomas, Radcliffe vessel - **ANNE THOMAS** (1882/1,418grt).*

The business prospered and four more vessels were acquired in the next three years, these, as became the custom of the company for their fleet, being registered in the names of single ship limited liability companies. Funding for this expansion was raised locally, much of this being from Welsh farmers.

A steady trade was built up carrying coal to Mediterranean ports, with homeward cargoes of grain loaded in Black Sea ports such as Odessa, and this form of trading was important to the company up until 1939. By the turn of the century the fleet had increased to 24 vessels, most of these being built by Ropners at Stockton on Tees and crewed by seamen from the Cardigan Bay area. By 1914, the company was the largest shipping firm in Cardiff with 28 vessels. At one point, the company had as many as 31 single-ship companies registered in its name.

Large platter made for the company by Hollinshead & Kirkham (1876-1900).

House flag as depicted on the platter and an old egg cup showing the correct colour of the flag.

Several vessels were lost during the First World War and the fleet consisted of only nine tramps when the war ceased. With compensation received for war losses four tramps were added to the fleet in 1925, ordering of these vessels being postponed until trading conditions improved. A number of vessels were laid up during the deep depression of the 1920/30s, all ships being returned to service by 1935.

During the Second World War many ships were lost as a result of the conflict and only five vessels came through the war unscathed. These were the **LLANBERIS**, **LLANGOLLEN**, **PETERSTON**, **FLIMSTON** and **LLANDAFF**.

***LLANDAFF** (1937/4,826grt) that survived the war.*

After the Second World War the port of Cardiff went into decline as fewer cargoes of coal were exported and this led to the demise of many local ship owners. Evan Thomas, Radcliffe & Company was one of those to suffer and over the next years only a few vessels were added to the fleet. This number included five tankers including one chemical tanker and for a time these ships became the main money earners of the company.

However by 1982 all ships had been sold and apart from briefly operating two small coasters for a year or so, the company ceased to exist.

GALBRAITH, PEMBROKE & CO, LONDON

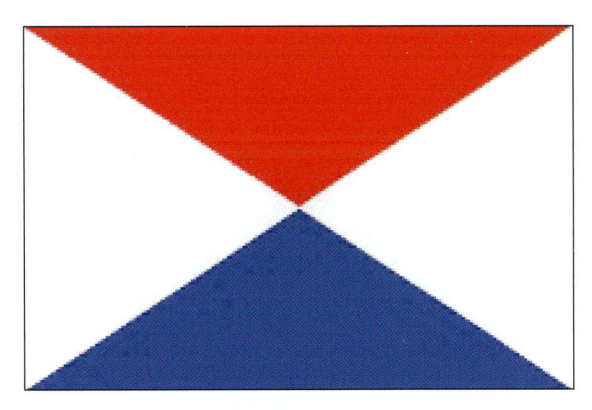

The houseflag of Galbraith, Pembroke & Co.

The company was formed in 1877 to trade mostly to the Mediterranean, previously being named Galbraith, Stringer, Pembroke & Co. The company vessels were for the most part general cargo ship/tramps, although a few tankers were briefly owned between 1895 and 1900. The tramp fleet expanded rapidly and in 1897 was registered under the ownership of Austin Friars Steamship Co, one of the company vessels, the **AUSTIN FRIARS** (1868/1,394grt) bearing this name. By this time the company was trading worldwide and by 1914 owned thirteen ships, but lost three to torpedoes during the First World War.

Large soup plate showing a registration mark for 1895 - manufacturer unknown.

In 1919 the fleet was sold to Houlder, Middleton & Co who traded the Austin Friars SS Co until 1921 when it went out of business. Galbraith, Pembroke & Co withdrew from ship owning during the inter war years but continued as ship brokers until 1940, when they purchased three old ships. The Basra Steam Shipping Co was formed in 1945 and operated until 1952 when it was sold to Graig Shipping Co, Cardiff. Galbraith, Pembroke & Co then returned to ship broking and ceased shipowning.

GEORGE GIBSON & CO LTD

George Gibson was born at Leith in 1758 and by 1797 he was a ship's agent and broker. In 1816, he became the manager of the Leith, Hamburgh & Rotterdam Shipping Company of sailing ships that ran to near Continent ports, becoming a shipowner in 1820 when he acquired the small vessel, **ISOBELLA**. The Leith, Hamburgh & Rotterdam Shipping Co was dissolved in 1844 and George Gibson purchased two sailing ships from the defunct company. In 1850 the first steamer was acquired for his company, **BALMORAL** (245grt), this vessel being followed by another nine steamers. A much larger steamship, the **ABBOTSFORD**, (1,035grt) was acquired in 1870 and the name of this vessel set the style of naming for many subsequent vessels this being names associated with the places, novels and characters in Sir Walter Scott's novels. Prior to this George Gibson had died in 1855 and his son Mungo Campbell took over the company over, and provided regular sailings between Scotland and the Netherlands, Belgium and France, mainly carrying cargo and a few passengers.

The company houseflag as depicted on its china.

After 1869, the company became involved in ventures trading with the Congo and carrying tea from the Far East, but withdrew from this trade in the 1880s. The company's larger steamers, such as the **ABBOTSFORD**, traded to the Far East. In 1883 the company commenced carrying a larger number of passengers (mostly on a seasonal basis) on the service from Leith to Rotterdam when they purchased the **ANGLIA** (1863/827grt), previously owned by the Dundee, Perth & London Shipping Company. Mungo Campbell Gibson died in 1890, being succeeded by his son, Campbell Gibson.

At the time of the commencement of the First World War in 1914 George Gibson & Company employed twelve passenger/cargo vessels on their service to Continental ports, six vessels being lost as a result of enemy action (mines or torpedoes) and two to collisions and one to grounding, during the war. Three company vessels were selected as 'Q' ships by the Admiralty (decoys), the **PEVERIL** being torpedoed and sunk in 1917. Campbell Gibson died in 1915 and as the grandson had been killed in action in 1915 there were no family members left to run the firm, which was acquired by the Somerville family of Edinburgh. The company became a limited liability concern in August 1916.

Comport supplied by A Sneddon & Son Ltd, Glasgow.

After the cessation of hostilities in 1918 the ships that had been lost in the conflict were replaced and within a couple of years the fleet numbered some seventeen ships. In 1920 the company amalgamated with another Scottish line that had provided competition on their traditional trade to the Netherlands and near-continental ports, the Rankine Line. This company had been founded in 1837 by James Rankine of Leith and following the agreement the joint companies traded as Gibson-Rankine Line. Even before this agreement came into being there had been a long understanding between the two lines to rationalise sailings in order to reduce competition.

During the late 1920s the number of passengers dropped considerably and as a consequence the number of berths on the vessels was either reduced to twelve passengers, or passenger accommodation removed completely, the lines being more concerned with cargo on the North Sea trades rather than passengers. The financial situation was worsened by the approaching Great Depression that affected the western world and the USA in the 1930s.

During the Second World War the company's vessels were requisitioned by the Government and operated as colliers, stone carriers, ammunition carriers and petrol carriers. Nine vessels were lost during the period of hostilities through a variety of causes, mines, U-boats, bombing, e-boats and collisions. After the Second World War the Gibson Rankine Line extended their services to Portugal and France with coal cargoes outbound and cork from Portugal homewards. The larger vessels traded even further afield as far as Canada and the Great Lakes, where they loaded grain. In the 1960s they moved into the carriage of liquid gas, these cargoes forming a large part of the company's business. These vessels were operated within the Unigas International pool, including Dutch and German owned ships and the vessels were operated under the name of Gibson Gas Carriers Ltd.

In 1972 the company had been sold by the Somerville family to the Runciman family and all vessels were registered under the name of Anchor Gas Tankers Ltd, the Anchor Line being one of the shipping companies owned by the Runciman Line, with George Gibson & Co managing the company's gas tankers. The Gibson part of the business was sold in 1986.

GEORGE MILNE & COMPANY, ABERDEEN
(INVER LINE)

George Milne & Co were one of the last British ship owners to operate sailing vessels during the first two decades of the 20th century, carrying such cargoes as grain and coal over long ocean passages, many to Australia. The company had owned a number of small wooden vessels during the latter part of the 19th century. In 1889 they had the steel three masted barque **INVERURIE** built, the first of a number of ships bearing the prefix "INVER" and the first of fifteen similar barques. She had a woman as a figurehead and had a registered tonnage of 1309 gross tons.

Inver Line tea cup and saucer (manufacturer unknown).

For the most part the barques had a gross registered tonnage of between 1300 and 1500 tons and all were built in Scotland. One authority - Basil Lubbock - stated:

"The INVERS were very pretty little vessels, painted French grey with a thin white line running along the sheet strake, and with white figureheads, and their yards and masts painted mast colour. Being very well looked after and extremely well found and victualled, they were very popular amongst seamen".

The final vessel to be built for the Inver Line was the **INVERNESS** built in 1902 by McMillan & Son. She was somewhat larger than the rest of the fleet registering 1959 gross tons. In 1918 she was lost to fire and abandoned at sea and soon afterwards George Milne sold his few remaining ships and retired.

*A typical Inver Line barque, the **INVERNEILL** sold when the fleet was sold to Sir William Garthwaite for £13,000. She was renamed **GARTHNEILL**. Her last passage, after many notable voyages to Australia, was made in 1925.*

GREAT WESTERN STEAMSHIP COMPANY

The Great Western Steamship Company was founded in Bristol by Mark Whitwill & Son to operate passenger and cargo services from Bristol to New York. The inaugural sailing was made by the iron screw steamer **ARRAGON** (1,317 tons) which sailed on 1 July 1871. This vessel was joined by the **GREAT WESTERN** (1,541 tons) in 1872, built specially for the service (although this vessel was also used from 1875 on voyages from New York to the Mediterranean). Disaster struck this vessel as she stranded on Long Island in 1876 whilst on a voyage from Messina in Sicily to New York and became a total loss. Additional vessels were chartered for the New York route and two new steamers were commissioned in 1875 from Messrs Richardsons of Stockton - **CORNWALL** (1,878 tons) and **SOMERSET** (1,923 tons).

Five further passenger/cargo vessels were brought into service over the next few years (three being built for other owners and two for the Great Western Steamship Company). However, passenger numbers decreased and no further passengers were carried after 1885.

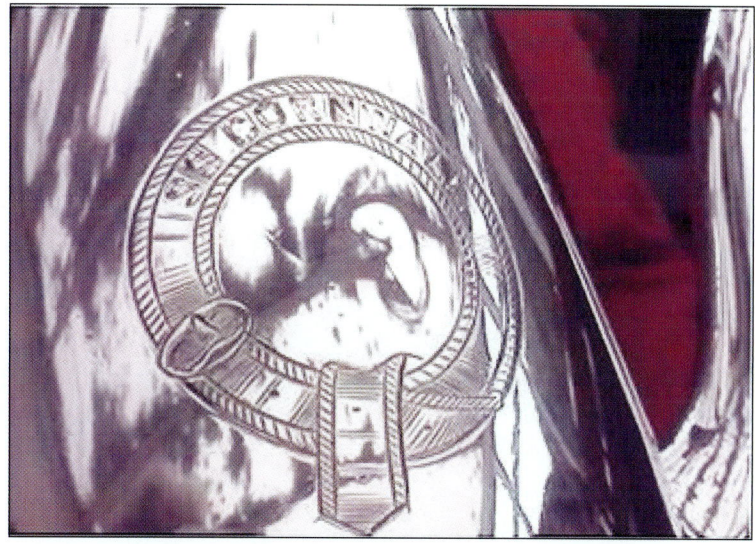

*Elegant silver plate coffee pot made for the **CORNWALL**.*

The **CORNWALL** was sold in 1886 to Turkish owners and became the **HASSAN PASHA**.

Cargo services were commenced from the new docks at Avonmouth to Montreal in 1879 and cargo services continued to New York. These were continued until 1895, by which time the fleet was reduced to one cargo steamer - the **MONMOUTH** - and the company was wound up the same year. The main factor in the demise of the company was the competition on the Bristol - New York service by the Bristol City Line of Charles Hill & Sons which commenced in 1878. This company only carried cargo and was not faced with the great expense of building and maintaining passenger ships.

GULF TRANSPORT LINE
(J H WELSFORD & CO, LIVERPOOL)

The Gulf Transport Line was owned and managed by J H Welsford & Co, Liverpool and registered as a company in 1902 although the firm had been in business since 1887. Included in the business were several iron sailing ships that had been owned previously by the Leyland Line. These were taken over from R W Leyland in 1909, competition from steamships creating poor trading conditions for sailing vessels.

In the early 1900s the Gulf Transport Line operated seven modern steamers on a regular service from Liverpool to New Orleans and Galveston on the Gulf of Mexico, USA, hence the name of the company. Services were also offered to other parts of the world depending on what cargo was available.

China platter, showing the company house flag and name. Cauldon Ware registered number Rd. 171711. The suppliers were D A Nesbitt & Co, Liverpool. This pattern of china was used by a number of Liverpool based companies such as the British Shipowners Co.

Small bowl showing the name of the Union Steamship Co of British Columbia.

Supplier's mark.

In 1911, J H Welsford and Company purchased controlling interests in the Union Steamship Company of British Columbia. This company had operated five steamers between Vancouver, the terminus of the Canadian Pacific Railway and to Port Simpson, the terminus of the Grand Trunk Pacific Railway.

Under this new management, the Union company entered the day excursion and resort business along the western coast of Canada by offering passenger services and then building and operating several resorts on Bowen Island, the Sechelt peninsula and at Whytecliff, Canada.

The Union Steamship Company was British owned until 1937.

HAIN STEAMSHIP CO LTD

The small Cornish fishing village of St Ives was the birth place of one of the most enduring fleets of British tramp steamers/cargo liners, although in the latter years the company was managed from Cardiff. Several generations of male Hains were christened with the name 'Edward', and several were Master Mariners. The family interest in ship owning commenced in 1816, when Edward Hain took a part ownership in the fishing lugger **DASHER**. Between then and 1901, when the Hain Steamship Company was founded, a total of twelve sailings ships - a mixture of schooners, brigantines and barquentines - were owned by the Hains, often with the financial backing of the Penzance merchants and bankers, Thomas & William Bolitho. They were also friends of the family.

Edward Hain, the fourth of the name, was the driving force behind the family becoming steamship owners and the first steam ship to be owned was named **TREWIDDEN**, after the Bolitho family estate near Penzance. She was built in 1878 by J Redhead & Co, South Shields and was the first of a total of eighty seven ships built by Redheads for the company, an amazing relationship. Her name set the pattern for the names of all subsequent Hain Line ships which commenced with the prefix TRE... Initially the vessels were owned by single ship limited liability companies until the Hain Steamship Company was registered on 16 September 1901.

Pint mug manufactured by Pountney, Bristol, showing the company flag used between 1878 and 1937.

Prior to the First World War the Black Sea grain trade was important to the company and after the war started the **TREVORIAN** (4144grt) was trapped in the Black Sea and was seized by the Russians. Two vessels, the **TREGLISSON** and **TREVIDER** were in German ports and they were detained throughout the war during which the company lost a further eighteen vessels, with considerable loss of life. Apart from the loss of seafarers, Edward Hain (now Sir Edward) lost his son when he was killed at Gallipoli in the Dardanelles in 1915 serving as a Captain in the Royal 1st Devon Yeomanry. Sir Edward never recovered from this and he died in 1917.

A pint mug manufactured by Ashworth Brothers circa 1948-1965.

On 27 October 1917, the P&O Steam Navigation Company made an offer for the Hain Steamship Company, which was accepted. Control of the firm passed to P&O, which then sold 50% of the company to their subsidiary, British India Steam Navigation Co, although management of the Hain Steamship Company remained with the Hain company. The same year, P&O bought the majority of the shares of the Mercantile Steam Ship Company, London and management of this fleet was entrusted to Hains. In 1923, all Mercantile Steam Ship vessels were transferred to the Hain Line and painted in their livery. However, these ships were not given Cornish TRE… names until 1936.

P&O's thirst for acquisitions continued and in 1923 it was decided that the Hain Steamship Company should purchase the shares in the four companies previously owned by Frank C Strick (the Strick Lines). The vessels of these companies consisted of passenger cargo liners, for the most part employed on routes to the Persian Gulf and Red Sea and had since 1920 been owned by Lord Inchcape's company, Gray, Dawes & Co. In 1928 Frank Strick re-acquired a minority 49% share in the Strick vessels.

A number of Hain Line ships were employed on full time charters to P&O and others built specifically to work under the P&O houseflag. This was very useful as the poor trading conditions in the 1930s took hold and many ships were laid up due to lack of trade including Hain Line ships which were laid up in Cornwall on the River Fal, close to Falmouth, an area still used for vessels that are laid up in hard economic times.

In the late 1930s trading conditions improved and by 1939 the Hain Line fleet consisted of 24 ships with a further three on order. The Second World War was disastrous for the Hain Line and all 24 ships, plus two of those on order, were lost, although a number of replacement vessels were obtained and at the termination of hostilities the fleet consisted of eleven motorships.

Egg cup showing the flag used between 1948-1965.

In the ensuing years, trading conditions fluctuated and by 1957 were in decline, this situation continuing into the 1960s. Another P&O subsidiary company was the Nourse Line and a joint management company was set up in 1964. In 1965 the Hain Steamship Company was renamed Hain Nourse Limited. Further reorganisation of the whole group took place in 1971 when a number of divisions were created. This included the P&O General Cargo Division and the remaining Hain Nourse vessels were transferred to this division, these being given names beginning with STRATH, for example **STRATHTEVIOT**, and pale blue funnels with a white P&O logo.

Although the remnants of the Hain Steamship Company continued for a few more years, this effectively was the end of the line.

*The **TREGLISSON** (5,970grt/1950) a typical Hain Line ship.*

J & C HARRISON LTD

The genesis of the J & C Harrison fleet was based on the coal trade from the north-east coast of England. James and Charles Harrison commenced as coal merchants and in 1887 decided to acquire their own colliers (many coal firms taking a similar course of action at this time). Their first two steamers were the **JAMES SOUTHERN** (847grt), later renamed **HARLOW**, and **CANNEL** (796grt), both acquired from the London firm of Harris & Dixon. The name of **HARLOW** was of special significance as the names of most of the company's subsequent vessels commenced with the prefix HAR...

The line's first new build was the **HARLYN** built in 1891 and over the next few years further ships were added to the fleet, these ships being employed on both the coastal and foreign trades. However, in 1896 the decision was made to sell their coastal collier fleet to William Cory & Son Ltd, a leading coal merchant and shipowner.

Three medium-sized ocean tramps were retained, the **HARDEN**, **HARCALO** and **HARPENDEN**. New buildings were quickly added to the fleet and by 1899 the company owned nine steamers. Ownership of these vessels was short lived as the commencement of the Boer War in South Africa saw a rapid increase in the value of ships and all the J & C Harrison fleet was disposed of by the end of 1899, at a profit. Throughout the years it was always company policy to acquire ships when they were available cheaply and sell them when a substantial profit was to be made. The director behind this policy was Charles Harrison, the company remaining a family business throughout its history.

Company cup and saucer manufactured by Bristol.

By 1904 the company resumed their ship owning activities, with ships being built or bought as they became available at an attractive price, the **HARCALO** (2) being the first of the new acquisitions. During the next few years ships were both bought and sold and by 1914 the fleet consisted of 13 large modern deep sea vessels, trading on a worldwide basis. By the end of the First World War in 1918 the fleet had reduced to only two vessels, either the result of submarine activity or profitable disposals.

Trading conditions in the 1920s and 1930s were very depressed. However, by judicious buying and selling the Harrison fleet by 1930 had increased to eight new tramps. A further eighteen new ships were delivered between 1932 and 1935. In addition, in the late 1920s the company took over the management of the London concern, Fisher, Alimonda & Co Ltd which had fallen on hard times. This company had absorbed an ex-North Atlantic liner company in 1915 - the National Steamship Company. Apart from this, Harrisons formed a new company in 1935, the Gowland Steamship Co Ltd and commenced a cargo liner service to the River Plate from the Baltic. At this time six of the vessels were registered to J & C Harrison, thirteen to the National Steamship Co and five to the Willis SS Company. A further vessel was registered to the Gowland SS Co.

The company entered the Second World War with a fleet of 25 handsome cargo ships, which was largely decimated in the war. A total of 19 ships were lost, plus another that became a constructive total loss and at the end of the war the company owned just five ships.

After the war the fleet was gradually rebuilt with a mixture of new ships (both motor and steam) and war standard vessels, but the company never reached the number of ships that it had owned pre-war. Trading conditions had changed and many countries were running their own national fleets and whilst the firm survived until 1979 it had become a shadow of its former self, the three remaining vessels being sold that year. Two of these were bulk carriers, **HARFLEET** and **HARFLEUR** and the third a coaster - **HARCOURT**.

*Model of the **HARCALO** - a typical J & C Harrison Ltd vessel.*

INCH LINE
(HAMILTON, FRASER & COMPANY, LIVERPOOL)

The Inch Shipping Co Ltd was formed by L Hamilton in 1879. Prior to this Hamilton owned sailing ships, but had decided to become a steamship owner. The first Inch Line vessel was the steamer **INCHCLUTHA** (1984grt). More steamers were acquired and by 1904 the Inch Line owned three ships, a further ship being owned under the name of Hamilton, Fraser and four under the name of the Rover Shipping Company.

*Company crest shown on their china, used on the **INCHKEITH**.*

Platter manufactured circa 1890.

The Inch Line vessels bore names such as **INCHULVA**, **INCHMAREE** and **INVERKEITH**. These vessels were employed on trades all over the world, including time charters and were in the size range of 3,750grt/4,800grt. The last Hamilton, Fraser vessel was sold in 1909.

INDO-CHINA STEAM NAVIGATION CO

The great Far Eastern trading company of Jardine, Matheson & Co was founded in 1832 by two Scotsmen, William Jardine and James Matheson. It rapidly became a powerful force in the Far East, trading in opium, cotton, tea, and silk. They were also substantial clipper and sailing ship owners. Their most famous vessel in the opium trade was the clipper **FALCON** and they also owned such fine ships as the tea clipper **STORNOWAY**. Early in their history the company operated in Canton and from 1844 the company head office was established in the British colony of Hong Kong, extending trade all along the Chinese coast. In the 1850s the firm commenced operating steam-powered vessels between India and China and also operating services along the coast of China.

In 1873 the company started the China Coastal Steam Navigation Co, which ran services between Chinese ports and Japan and eight years later in 1881 the British company of Indo-China Steam Navigation Co to operate the firm's river and coastal services. The river services to the lower, middle and upper Yangtze River became a very important sector of the company business and was a great rival of the other great British Trading company, also based in Hong Kong, Butterfield & Swire which owned the China Steam Navigation Company. Indo-China Steam Navigation Company foreign going ships were also employed on routes to Singapore, Japan, India and far eastern Russia and, from 1885 Hong Kong to the Philippines. The company ships bore Chinese names such as **FUH WO**.

At the turn of the century, over 50% of shipping on the Yangtze River was controlled by the Indo-China Steam Navigation Co and the China Steam Navigation Co and Shanghai became an important city for both of the rival companies. This was to change after the Second World War following a British-Chinese Treaty in 1943, which effectively shut the British companies out of the coastal and river trades and Shanghai. As a consequence of this, the Indo-China SN Co commenced a new service between China and Australia, later extending this to New Zealand. This route was from Hong Kong to Brisbane, Sydney, Melbourne, Wellington and Auckland outwards, with calls at Melbourne and Sydney homewards. Fine vessels were employed on this service bearing names commencing with the prefix EASTERN... for example **EASTERN ARGOSY** and **EASTERN TRADER**.

Small tureen marked with the company crest.

Services which included the carriage of passengers between the Far East and India were terminated in 1955. Both Indian and Japanese companies had taken over much of the trade to the Antipodes, the result of this being the closure of the British company of Indo-China Steam Navigation Company in 1974. However, this was not the end of Jardine Matheson's shipping interests and in 1985 this part of the company changed name to Jardine Ship Management Ltd.

Tea cup supplied by James Green & Nephew Ltd, London.

INDRA LINE
(THOS B ROYDEN & CO)

The oldest of the constituent companies that came together to form the Commonwealth and Dominion Line (Port Line) was Thos B Royden & Co. As early as 1800 Thos Royden was building wooden ships in Liverpool. He held shares in some of these ships and when trade was slack, some vessels were built and traded by the Royden family themselves. These vessels sometimes served on the Liverpool to India routes, but also to other parts of the world, including Australia. By the mid-1860s, iron sailing vessels were being built. Steamers followed a few years later, the first being **INDRA** (1888/3,528grt) which traded to India from 1888 to 1891. All the Royden steamers had names beginning with the letters 'IN'.

Platter from the Royden Line of Steamers.

From 1891 several of the steamships were chartered to G D Tyser for use on its New Zealand trade and were fitted with refrigeration machinery. The Royden shipyard closed in 1893 when the Mersey Docks & Harbour Board purchased the site for a dock extension.

Indra Line tea plate supplied by Stonier & Co Ltd, Liverpool.

Indra Line soup plate by George Jones & Sons/Stonier & Co Ltd, Liverpool.

The company formed the Indra Line in 1901 and the main routes served were from New York to the Far East. In 1914 three ships were contributed to the Commonwealth & Dominion Line (Port Line) and the Santa Clara SS Co was formed to operate a feeder service from the West Indies to New York and one ship, the **SANTA CLARA**, was built for this route.

The remaining New York service and the ships on it were sold to the Blue Funnel Line in 1915 and the Santa Clara SS Co remained Royden's only shipping interest. In 1920 a second ship, **PINAR DEL RIO** was built. However, the **SANTA CLARA** foundered in 1924 and in 1930 the **PINAR DEL RIO** was sold to the Bristol City Line and renamed **CITY OF MONTREAL**. This ended the shipowning business of the Royden family.

JAMAICA DIRECT FRUIT LINE/ JAMAICA BANANA STEAMSHIP COMPANY

In 1929, the Jamaica Banana Producers Association of Kingston, Jamaica (an organization of independent growers) decided to form their own shipping company to transport their fruit and the Jamaica Direct Fruit Line was created. The first four vessels owned by the company were refrigerated meat carriers, which had limited passenger accommodation. They were purchased from the Nelson Line and these HIGHLAND ships, each of about 7,500grt and built in 1910/1911, were renamed the *JAMAICA MERCHANT*, *JAMAICA PLANTER*, *JAMAICA PRODUCER*, and *JAMAICA SETTLER*. The four ships were converted for the carriage of fruit and management of the ships was entrusted to the Kaye Son & Co, London. In 1935 the company name was changed to the Jamaica Producers Steamship Company, three new vessels having been ordered in 1931 and built as the *JAMAICA PIONEER*, *JAMAICA PROGRESS* and the second *JAMAICA PRODUCER*, the first vessel of that name being lost to fire in 1933.

During World War Two, the British Admiralty commandeered all four ships owned by the company for war service, and they played an important role in delivering food to the United States and the United Kingdom. By the end of the war all but the *JAMAICA PRODUCER* were sunk. The *JAMAICA PRODUCER* also had the distinction of shooting down a German Stuka dive-bomber that attacked her in the English Channel, and for this, the Admiralty awarded her a commemorative plaque. She continued sailing until 1962 when she was taken out of service. Two replacement vessels were built by Lithgows in 1959 and 1962, *JAMAICA PLANTER* (3) and *JAMAICA PRODUCER* (3).

Comport for the Jamaica Direct Fruit Line manufactured by Royal Doulton circa 1929.

Sugar bowl made by Royal Doulton circa 1929.

Following that country's independence in 1962 the ships sailed under the Jamaican flag. The Jamaican Government took over the company in 1977 and a new concern, the Jamaica Producers Marketing Company took over the management from Kaye Son & Co and the Jamaica Merchant Marine Atlantic Line was formed jointly, with Transportación Marítima Mexicana SA and the Jamaican Government.

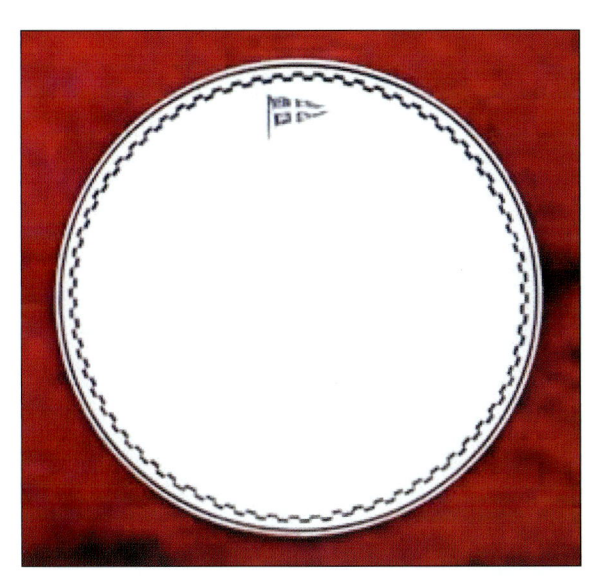

Dinner plate made by Royal Doulton and a tea cup made by John Maddock & Sons Ltd for the Jamaica Banana Producers Steamship Company.

The company flag was originally a red pennant with a white cross and the letters "JDFL" in the quarters. This was changed to a green pennant with a yellow cross and the new letters "JBPS", when the company name changed. The house flag was used as a decoration on the company china.

Funnel colour.

The JAMAICA MERCHANT ex HIGHLAND PIONEER in Liverpool.

JAPP & KIRBY, LIVERPOOL

Japp & Kirby owned and operated sailing vessels from Liverpool until 1882, after which they ceased owning vessels for a ten year period. In 1892 they resumed ship owning when they purchased four sailing ships and two steamers - the **OTTERSPOOL** (2878grt/1896) and **WILDERSPOOL** (2804grt/1894). These vessels were all sold by 1900.

A pair of company plates with floral decoration circa 1894.

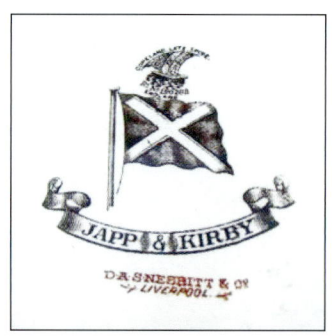

*Underside of the two plates showing the company houseflag,
the name of the manufacturer - Copeland late Spode and the supplier - D A S Nesbitt & Co, Liverpool
(the latter company supplied crockery to many leading Liverpool ship owners).*

In 1904 the company resumed ship owning when they had two steamers built, the **JURA** (3751grt/1904) and the **BARRA** (3761grt/1905). By 1911 the fleet totalled six steamers which were on occasion time chartered to other shipowners. These vessels were registered at Liverpool, although by now the company office had moved to London.

*Flo blue sandwich plate top marked with the company name and house flag in blue,
manufactured by W T Copeland.*

One of the senior family members/partner, Russell Japp, died in 1912 and the company ceased trading under the name of Japp & Kirby the same year.

However, the company was re-established in 1916 by the other partners under the name Japp, Hatch & Co, London, which managed the Mondrich SS Co and later the County SS Co, and Oxford SS Co Ltd until 1929, when the economic situation was such that the company had to close down and cease trading.

JOHNSTON LINE
(WILLIAM & EDMUND JOHNSTON, LIVERPOOL)

In 1872 William & Edmund Johnston purchased a small steamer, the **PLYNLYMMON**, to operate sailings to the Mediterranean, Black Sea, Greece and Turkey, further vessels soon being purchased. In 1874 the company bought the steamer **ARDMORE**, which set the system of naming the majority of their subsequent vessels with the suffix ...MORE. In 1880 the company joined forces with an American railroad company, the Baltimore & Ohio, to operate transatlantic sailings to Liverpool, live cattle being an important part of this trade. Ten years later similar sailings were commenced from London to Boston. In 1894 a 'pooling' arrangement was agreed with Furness Withy & Company and a Liverpool to Montreal service commenced.

In 1914 Furness Withy purchased a 50% interest in the Johnston Line, full ownership by Furness Withy being completed in 1916. At the time of the initial ownership in 1914 the Johnston fleet consisted of 18 ships, four of these being large vessels employed on the North Atlantic and fourteen smaller ones on routes to the Mediterranean and Black Sea. The whole fleet was more or less decimated during the 1914-1918 conflict, only the **CRANMORE** being in service at the end of the conflict.

During 1934 Furness Withy decided to merge their Liverpool fleets (the only Johnston Line services surviving the depression being those to the Mediterranean and Black Sea) and the Johnston Line fleet was merged with the Neptune Steam Navigation Company and the Warren Line. The latter company operated services from Liverpool to St John's, Newfoundland - Halifax, Nova Scotia and Boston.

China soup plate registration number Rd.171711
Cauldon Ware supplied by D A S Nesbitt & Co, Liverpool Circa 1894.
This pattern of china was supplied to several Liverpool-based companies by Nesbitt and by
Thos F Bennett & Co another Liverpool supplier. The platter in the same pattern was supplied by Bennett.

China egg cup - manufacturer unknown.

The Johnston Warren Lines was formed in December 1934. The former Neptune Line vessels bore the prefix LONDON, for example **LONDON EXCHANGE**, and the Warren Line ships were named **NEWFOUNDLAND** and **NOVA SCOTIA**. The names of the latter ships were given to new vessels under Furness Withy ownerships in the 1950s.

KHEDIVIAL MAIL LINE

It is thought that the Khedivial Mail Line was founded in Egypt in 1858 although details are uncertain and even the name of the early company is subject to speculation. However, it is known that the Khedivial Mail Steamship & Graving Dock Co was formed in 1898 to operate ships and docks owned by various departments of the Egyptian Government. The company fleet was registered under the British flag and operated passenger and cargo services between Alexandria, Constantinople and Syrian ports and between Suez and Red Sea ports. Later services were extended to Piraeus, Malta, Marseille and Cyprus.

Old dinner plate. The inscription in ancient Arabic reads "Pasha's Postal Service".

The line became part of the great Peninsular & Oriental Steam Navigation Co in 1919 and continued under their ownership until 1924. In 1936 the company was re-formed in Alexandria as the Pharaonic Mail Line until 1941, when the company name reverted to the Khedivial Mail Line.

An attractive platter bearing the Khedivial Mail Line mark on the underside.

For the most part the Khedivial Mail Line concentrated on routes within the Mediterranean until March 1948, when a transatlantic service was commenced between Alexandria, Naples, Boston and New York. The first vessel employed on this route was the **KHEDIVE ISMAIL** (1944/8,193grt), a former American-built Victory ship. She was followed by a second similar former Victory ship, **MOHAMED ALI EL KEBIR**. Apart from Naples call were sometimes made at Genoa, as well as at Beirut and Marseilles and in the USA, occasionally at Baltimore, Charleston and Philadelphia. A third vessel was acquired for service on the route, the former Canadian National Steamships **LADY NELSON** renamed **GUMHURYAT MISR**. Following this acquisition the homewards bound steamers after February 1954 called at Alexandria, thence proceeded through the Suez Canal to Bombay and Karachi. This short lived extension proved to be unviable and only lasted for a few months.

Another example of a crest on the company china and company advertising poster.

In July 1961 all Egyptian shipping lines were nationalised by the Egyptian Government and the Khedivial Mail Line became part of United Arab Maritime Co who discontinued New York sailings, but for a time until 1965 made passenger and cargo sailings to Canada.

The ancient Arabic script shown on the first plate illustrated took some time to identify and it was only after I took the plate to the Anglo-Egyptian Trade Commission in London, who had a member of staff who could read ancient Arabic, that I was able to confirm it to be the crest of the Khedivial Mail Line and its provenance.

KNIGHT LINE
(GREENSHIELDS, COWIE & CO, LIVERPOOL)

Houseflag of the Knight Line.

Greenshields, Cowie & Co, was created in 1877 by the merger of C G Cowie, Son & Co, which owned a fleet of sailing ships and Greenshields & Co, Liverpool. The firm was for the most part engaged in the Indian cotton trade to the UK but later also traded to the Far East, Australia and New Zealand. The American Civil War created problems for the company. However, after this war ceased, the company returned to the cotton trade from the USA.

The firm owned a total of eleven iron sailing ships, the first to be given a "Knight" name was the **KNIGHT OF THE GARTER**. She was built in 1877 and was of 1,494grt. In 1881 the company took delivery of their first steamship, the **KNIGHT OF ST PATRICK** (1881/2,249grt) and their last sailing ship was sold in 1897.

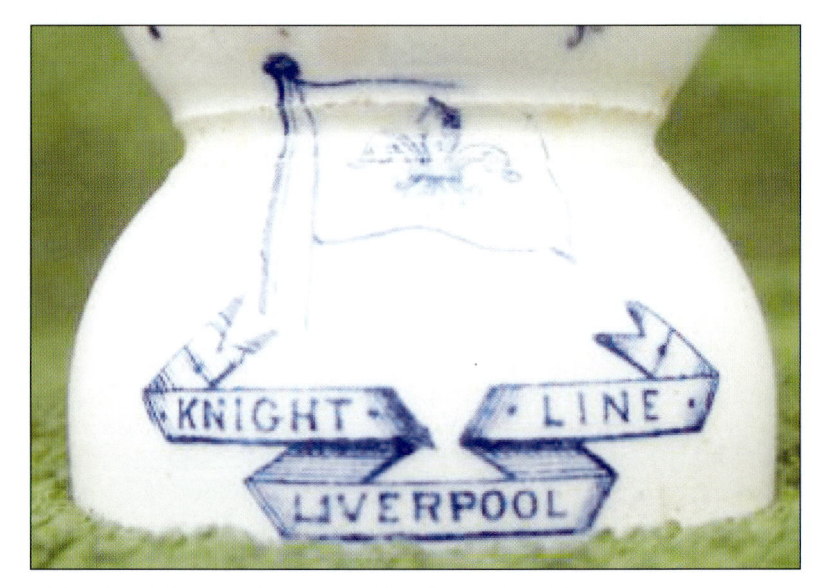

Company egg cup manufactured by Cauldon and supplied by D A S Nesbitt.

By 1914 the company owned four ships which were chartered to Alfred Holt & Co (the Blue Funnel Line) and in 1917 the fleet was sold to Alfred Holt. In 2014 the company still survived as shipping and forwarding agents, but no longer owned any ships.

LANCASHIRE SHIPPING COMPANY

(JAMES CHAMBERS & CO)

The Lancashire Shipping Company was founded by James Chambers in 1896 to take over the assets of the Lancaster Shipping Company, which could trace its history back to 1849 when two Liverpool shipbrokers, John Pilkington and Henry Threlfall Wilson, set up the White Star Line of Boston Packets. James Chambers, also a shipbroker, was a major shareholder in the line until 1865 when he left the firm. This company collapsed a couple of years later.

In 1865 the Lancaster Shipping Company was formed by Wilson & Chambers, as managers, and in 1876 the company bought the Greenock-owned barque **WARWICK CASTLE** that had originally been built for Donald Currie. The name of this vessel set the pattern of names for all subsequent vessels owned by the company, all being named after castles in Lancashire or north western England and registered in Lancaster. The steamer **EGREMONT CASTLE** was delivered in 1890, the first to be owned by the company. The last sailing vessel to be owned by the Lancashire Shipping Company was the **WRAY CASTLE** of 1889 which was disposed of in 1901. By this time the fleet consisted of steamers employed as tramps, trading wherever cargoes became available.

Company chamber pot and egg cup - Cauldon circa 1896.

At the outbreak of the First World War in 1914 the fleet consisted of fourteen vessels, two of which were lost in the conflict. Tramping was resumed after the war and several vessels were also chartered to cargo liner companies, such as the Barber Steamship Lines.

The Second World War proved to be disastrous for the company and by 1943 half the fleet had been sunk. The decision was taken to sell the remaining ships and the company itself was sold to a Hong Kong company, Barber Steamship Lines. James Chambers & Company withdrew from ship management in 1946 and the management was taken over by Moller & Company, Hong Kong.

Mollers continued to operate vessels under the name of the Lancashire Shipping Company until 1950 when their two remaining vessels were sold, this marking the end of the company.

MOGUL LINE LTD

The Mogul Line was founded in 1877 as the Bombay & Persia Steam Navigation Company. The managers of the company were Messrs Turner, Morrison & Company of Liverpool, Calcutta and Bombay, who were involved in the pilgrim trade to Mecca for many years. Eventually the company was taken over in 1912 by a combination of the British India Steam Navigation Company and the Asiatic Steam Navigation Company - both actively involved in the Indian Coastal trades between Calcutta and Bombay. The Asiatic Steam Navigation Company also held the Indian Government mail contract between Calcutta and the Andaman Islands.

*Milk jug for the Mogul Line Limited,
made by Messrs Dunn Bennett & Co, supplied by McSymons & Potter circa 1955.*

In 1939 the Bombay and Persia Steam Navigation Co was renamed the Mogul Line and in 1960 the line became entirely Indian owned. In 1963 the Mogul Line was taken over by the Shipping Corporation of India.

JOHN MORISON & CO

John Morison & Co was a London-based sailing ship owner during the latter part of the 19th century. However, little information has come to hand regarding the history of the company.

It is believed that the soup plate illustrated above could have been used on vessels such as the composite ship **FIERY CROSS**. This famous sailing ship was built in 1860 at Liverpool by Chaloner for John Campbell of Glasgow. Her registered tonnage was 702.31. She was bought by John Morison in 1874 and continued trading to the East. In 1877 she was sold to other London owners, finally being sold the same year to Norwegian owners as **ELLEN LINES**. Two years later she experienced a fire in her cargo of coal and sank in the Medway.

Wm Fairbairns, who either supplied or manufactured the china, is not recorded in the standard works of pottery marks.

MOSSGIEL STEAMSHIP CO LIMITED
(J BRUCE & CO, GLASGOW - MANAGERS)

Company crest and house flag.

Mossgiel Steamship Co Ltd was incorporated about 1894 as a shipowning firm in Glasgow under the management of J Bruce & Company which operated steamships from 1892 under the name of Bruce, Babtie & Co. The company traded goods into Mediterranean ports and built up its own fleet of steamers to undertake the trade. Regular services were provided from Glasgow to ports in Italy and Sicily, Marseilles (France), Alexandria (Egypt) and to various Spanish Mediterranean ports, these being served by small general cargo vessels such as **ALCOR** (1919/1,400grt), **ALMENARA** (1922/1,900grt), **ALHAMA** (1938/1,400grt) and **ALPERA** (1920/1,800grt). These vessels carried up to six passengers.

Soup plate supplied by A Sneddon & Sons, Glasgow.

In 1958, the company's directors agreed to an amalgamation with the Ellerman group of companies and the company became a subsidiary of Ellerman & Papayanni Lines Ltd which already operated vessels on similar routes to the Mediterranean out of Liverpool. By this time the Mossgiel Steamship Company was operating two vessels.

Two Ellerman & Papayanni ships were transferred to/managed by J Bruce & Company, the **CROSBIAN** built by William Gray & Company (1947/1,518grt) being transferred in 1963 and the **CORTIAN** (1962/537grt) in 1968. The company was dissolved in 1994.

*The **ALHAMA**.*

NEW YORK & SOUTH AMERICA LINE
(CHAS G DUNN & CO LTD, LIVERPOOL)

The Liverpool firm of Chas G Dunn & Company evolved from the Globe Shipping Company of 1887, which had started as Herron, Dunn & Company in 1882. From the beginning their ships bore names with the suffix of HALL eg the **MISTLEY HALL** (1874/1,867grt).

Coffee can and saucer showing the company house flag - Cauldon Ware supplied by D A S Nesbitt & Co Liverpool.

In 1903 the American US Steel Products Company obtained a contract with the Government of Chile for railway equipment and consequently bought a controlling interest in Charles G Dunn & Company, Liverpool, which was to manage the New York & South America Line vessels on the service. These were to be registered under the British flag (an early "flag of convenience") to reduce running costs which would have been much higher if the ships had been registered in the USA and given American crews. As trade to South America continued to expand, two Chas G Dunn vessels were placed on the new service to South America under time charter on behalf of the US Steel Products Company. These were operated as the New York & South America Line and in total five vessels were ordered by Chas G Dunn, which were eventually to be acquired by the US Steel Products Company in 1913.

These vessels were the:-

CHARLTON HALL	(1907/4,853 tons)
CRASTER HALL	(1909/4,319 tons)
CROFTON HALL	(1913/5,378 tons)
FOXTON HALL	(1902/4,247 tons)
HOWICK HALL	(1910/5,096 tons)

HOWICK HALL *(1910/5,096 tons) from a painting by Laurence Dunn.*

Oval bowl - Cauldon Ware supplied by D A S Nesbitt & Co Liverpool.
This pattern was used by many Liverpool based shipping companies such as the Dominion Line.

Apart from the Chas G Dunn owned vessels, the US Steel Products Company also bought five vessels operated by the London-based Isthmian Steamship Company under the British flag. Other vessels were chartered as required. With the outbreak of World War One all the British registered ships were requisitioned by the British Government. This brought about the end of the British-based US Steel Company ships.

US Steel Products Company later adopted the name of Isthmian Lines for its shipping interests, as shown on the coffee cup and saucer illustrated below.

Coffee can and saucer showing the Isthmian Steamship Lines house flag.
- Cauldon Ware supplied by D A S Nesbitt & Co Liverpool.

J NICHOLSON & CO, LIVERPOOL

The **MARY NICHOLSON** was a ship built in 1862 by Lumley, Kennedy & Co at Whitehaven. She had a wooden metal sheathed hull and measured 604 gross registered tons. Her length overall was 165.2 feet and she was classed A1 at Lloyds. Her port of registry was Liverpool.

She was originally to be named **PRINCESS ALEXANDRA** and launched as such but was renamed by her owners, J Nicholson, who employed her initially on their trade from Liverpool to China. Her maiden voyage from Whitehaven commenced on 24 December 1862. She sailed under the command of a Captain James Freebody and her next three years were spent in the Far East.

Dinner plate dated 1863 showing the name of the vessel and the Liverpool crest.

In October 1869 the ship was dismasted in a typhoon in the Kino Channel near Hiogo, some 240 miles south-west of Yokohama, whilst on passage from Yokohama to Hong Kong in ballast. She was abandoned by her crew in latitude 32 deg N and longitude 136 deg E and on 2 October she was boarded by the crew of the **CLYDEVALE**.

The Liverpool Mercury dated Thursday 25 November 1869 recorded:-

*"A few days ago we published the fact that the Liverpool ship **MARY NICHOLSON** had been passed derelict off the coast of Japan, but that nothing had been heard of her crew. A telegram has been received from her captain (Freebody), dated from San Francisco, and stating that the ship had been abandoned. The crew, with the exception of a seaman named Riley, were all saved ... The **MARY NICHOLSON** was the property of Mr John Nicholson, of Kirkdale, Witham, Wigtownshire, whose Liverpool agents are Messrs Nicholson and M'Gill. We believe the vessel is uninsured, Mr Nicholson not insuring 'on principle'. The ship is valued at £10,000".*

PALM LINE

William Lever was one of the great names in British commerce and controlled most of the soap industry. As a result of his desire to protect his supply of palm oil and palm kernel oil from West Africa predominantly Nigeria, he formed the Bromport Steamship Company in 1916 to carry his own produce. Four years later he bought the very old established trading organization - the Niger Company - which itself had commenced shipowning in 1928. This company was in direct competition with the African & Eastern Trade Corporation Limited which had entered shipowning in 1924. William Lever, by now Lord Leverhulme, died in 1925 and in 1928 the Niger Company and African & Eastern Trade Corporation merged to become the United Africa Co Ltd. At the same time Lever Brothers merged with the Dutch Margarine Union to form the famous Unilever Group.

Following the Second World War the United Africa Company decided that its shipping interest should become a separate company and act as a 'common carrier'. This meant that as well as company cargoes, the ships would carry any cargo that was available, for any shippers. Therefore, in February 1949 the Palm Line was created. This was achieved by reviving the dormant articles of association of the old Southern Whaling and Sealing Company, and changing the name to Palm Line. The Lever interest in whaling had commenced circa 1916, whale oil being used for the manufacture of both soap and margarine.

Salad plate manufactured by Grindley circa 1970.

At the time of inception the fleet consisted of fifteen vessels, all bearing names ending with the suffix ... IAN eg *ETHIOPIAN* (1936/5,424grt). She was renamed **BENIN PALM**. Likewise all other vessels were similarly renamed. The vessels were traditional general cargo ships and the fleet included two vegetable tankers. The length of all ships was limited to a length of 500 feet and a shallow draught so as to be able to navigate the shallow rivers and creeks common along the West African coast and to cross the sand bars at the entrance of the rivers. The ships were also provided with deep tanks to carry vegetable oil and capable of carrying a few passengers.

Apart from the Palm Line two other British shipping companies were heavily involved in the West African service, the long established Elder Dempster Lines and the Guinea Gulf Line (the shipping division of the West African trading company, J Holt & Company), both based in Liverpool.

Due to the economic conditions, port congestion and civil unrest in West Africa and the emergence of the Nigerian and Ghana state-owned shipping companies, Nigerian National Shipping Line and the Ghana Black Star Line, there was a serious decline in the availability of cargo, although Palm Line had been admitted to the conferences UK/West Africa Lines Joint Service (UKWAL) and Continent/West Africa Conference (COWAC).

The days of the Palm Line were numbered and in 1985 it was taken over by the Ocean Transport & Trading Company (Ocean Fleets/Blue Funnel Line), which already owned the Elder Dempster Lines and the Guinea Gulf Line. In 1989 the Ocean Trading company decided to withdraw from the West African trade and in 1989 disposed of the Elder Dempster Lines, Guinea Gulf Line and Palm Line to the French company Delmas Vieljeux. The Palm Line thus no longer existed.

PLANET LINE

The Leyland brothers commenced operating sailing ships in 1874, although the first vessel to be owned by R W Leyland & Company was the iron barque **DOXFORD**, which they purchased in 1875. The sailing ship fleet increased over the years and included the famous vessel, **WAVERTREE**.

By the end of the 1880s they were contemplating adding steamships to their fleet and in total the company was to own five PLANET steamships, all bearing names of planets with the prefix of PLANET before the name of the planet. The first two acquisitions in 1894 were the steel steamships **PLANET VENUS** (3,689grt) and the **PLANET MERCURY** (3,223grt). In the event the **PLANET VENUS** only sailed briefly for R W Leyland & Company, as she was sold in 1897 to J Holman & Sons.

Dinner plate showing the PLANET LINE flag and name.

Various attempts were made to operate the steamships on a service to the United States of America and a joint service between Philadelphia and Avonmouth was commenced in 1891 with Manchester Liners. Trading conditions were difficult, not being improved by a degree of mismanagement. This led to management of the company being taken over by J H Welsford & Company, including the remaining steamships **PLANET MARS**, **PLANET NEPTUNE** and **PLANET VENUS** (II).

The end of the company came in 1909 when J H Welsford & Company took over the shipping of interests of the Leyland Shipping Company in entirety.

RANKIN, GILMOUR & COMPANY LIMITED

Rankin, Gilmour & Company originated from a Glasgow-based timber importing firm dating from 1804. This company, Pollok, Gilmour & Company set up timber operations in Miramichi, Canada, and another branch, Robert Rankin & Company, in Saint John, New Brunswick. The company was very successful, owning sawmills and many wharves and a large number of Canadian-built sailing vessels with which they used to export their timber. At one time they operated some 120 vessels in the lumber trade. In the winter months many of these vessels were employed in the cotton trade from New Orleans and Mobile, Alabama.

In 1838, Robert Rankin who had moved to Canada, returned to Glasgow, the company becoming Rankin, Gilmore & Company and eventually set up its head office to Liverpool. In 1865 the company built a sailing ship called the **SAINT MAGNUS** and this name was the first vessel of the fleet to carry the prefix SAINT in the name and to the fleet eventually being known as the 'Saint Line'. In 1880 the line acquired its first steamer, the **SAINT COLUMBA** (2,223grt) and formed the British & Foreign SS Co. By the end of the century the company owned one sailing vessel and a dozen or so steamships, which later traded worldwide, including regular services from New York to China and Japan and from New York to the River Plate.

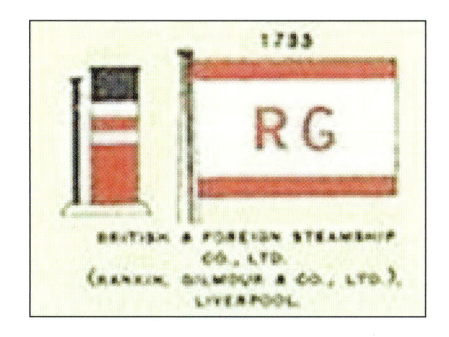

*Original house flag of Pollok, Gilmour & Company
and the later flag of Rankin, Gilmour & Company (British & Foreign Steamship Company).*

A china ladle made by Davenport & Co, used on board the steamer SAINT MARNOCK.

Davenport embossed mark.

The **SAINT MARNOCK** was built by James Laing, Deptford, Sunderland, in 1889 for the British & Foreign SS Co. She was a steel screw steamer of 2969grt. In 1899 she was sold to Wm Thomson & Co, Dundee, and then in 1908 to the Cairn Line of Steamers. She was finally sunk by a mine in 1915 whilst on a voyage to Montreal.

During the course of World War One the company lost three vessels to enemy action, two further ships becoming marine casualties. In 1917 the majority of the company's ships were sold to the well-known Liverpool shipping company, T & J Harrison (Charente SS Company), and the British & Foreign SS Co was wound up in 1918.

A new 'Saint Line Limited' was formed in 1919, the residue of the original fleet trading primarily to ports in the River Plate and South America from Europe. This was operated from the Liverpool office until 1931. At this time, operation of the vessels was transferred to an office in London and by the mid-1930s Rankin, Gilmour & Company no longer controlled the company and ownership passed to Mitchell Cotts & Company.

RED CROSS LINE OF STEAMERS TO NORTHERN PORTS OF BRAZIL
(R SINGLEHURST & CO, LIVERPOOL)

R Singlehurst & Company had owned and operated sailing vessels to northern Brazil ports for many years. In 1869 they decided to venture into steamships and formed the Red Cross Line of Steamers to Northern Ports of Brazil. The first three steamships were quite small and built by T Royden & Sons at Liverpool, all being approx 1,300 tons gross; they were **PARAENSE**, **MARANHENSE** and **CEARENSE**. The suffix ... ENSE became a system of naming used for many later Red Cross vessels.

Large platter depicting the line name and house flag.

The new line operated in direct competition with the Booth Line to most northern Brazilian ports, Para being one of the main ones served, although the Red Cross Line operated further up the River Amazon than Booths, to Manaus some 1000 miles up river from the open sea.

From 1870 agreement was reached between the two companies to share the trade, this agreement continuing until the demise of the Red Cross Line. A fortnightly service was instigated and regular calls were made at Portuguese ports, by both the Red Cross Line and Booth Line. This was not surprising considering the close links existing between Brazil and Portugal. Both companies also owned tugs and lighters on the Amazon to service their trades and to distribute cargoes.

In 1897, R Singlehurst & Company commenced operating direct services from Liverpool to the Peruvian port of Iquitos, some 2000 miles up the Amazon, forming the Red Cross Iquitos Steamship Co Ltd for this purpose. The Booth Line similarly created a new company, the Booth Iquitos Steamship Co Ltd. These two companies were amalgamated in 1901 and became the Iquitos Steamship Co Ltd, later being absorbed into the main Booth Line fleet in 1911.

RIO GRANDE DO SUL STEAM SHIP COMPANY LIMITED

Large company platter supplied by James Green & Nephew, Queen Victoria Street, St Pauls. This is an exceedingly rare item.

The Rio Grande do Sul Steam Ship Company, which was a very short lived company, was formed and registered on 19 February 1874 with the intention of running regular direct steamship services between London and Europe to the Brazilian province of San Pedro do Rio Grande do Sul. Up until this time sea communications had mostly been carried out by sailing vessels although high value cargoes were shipped by steamers to Rio de Janeiro where they were transhipped to local vessels. The head office of the new company was No 1 East India Avenue, London EC, and the main shareholders were Messrs J C Im Thurn & Co.

Initially it was proposed that three or four steamers be built by Messrs Wm Hamilton & Co (one ship) and Cunliffe & Dunlop (three ships), both situated at Port Glasgow. The latter firm specialised in low draught vessels, where the rivers and ports of destination had shallow water limitations. These vessels were to be capable of accommodating 24 1st class passengers and 300 steerage and to load a substantial amount of cargo, being classed 100 A1 Lloyds Special Survey. Their port of registry was London.

Sailings were to be monthly, alternately from Hamburg or Antwerp calling at Oporto and from London calling at Le Havre and Bordeaux, for cargo and passengers.

Supplier's mark.

Clearly the company's expectations of a successful operation were ill founded as an article in the 'Teesdale Mercury" states:-

> NOTICE is hereby given, that a petition for
> the winding up of the above-named Company
> by the Court of Chancery, was, on the 7th
> day of September, 1875, presented to the Lord
> Chancellor by John Conrad im Thurn.

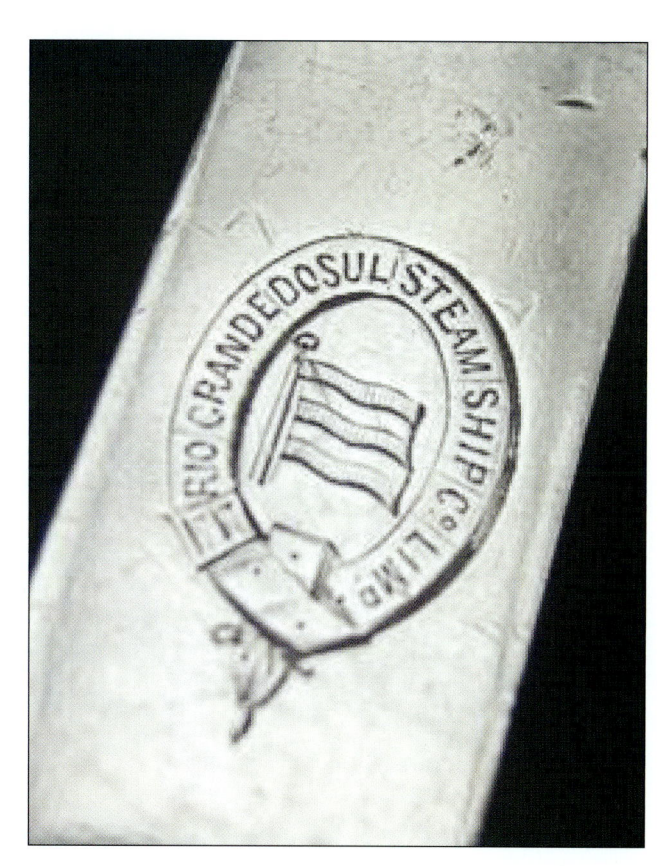

Company crest as depicted on their china and silver plate tableware.

Information regarding the company's vessel is scant and whether all four ships were completed and commissioned is not clear, however the **RIO GRANDE DO SUL** (1,135grt) was built in 1874 by Wm Hamilton and sold in 1876. After this she had various owners finally being a marine casualty, whilst she was owned by a Russian company, when she foundered off Saddle Island near Shanghai in September 1905.

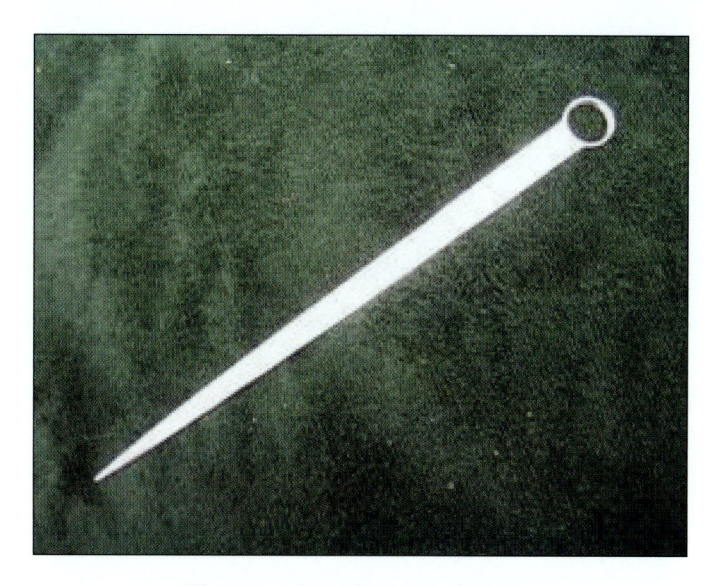

Electro plated meat skewer.

ROYAL EXCHANGE SHIPPING COMPANY LIMITED
(THE MONARCH LINE)

The company house flag.

Like many other shipping companies, the Royal Exchange Shipping Company started off as sailing ship owners, their early sailing ships being employed on the routes to the Mediterranean and Far East from the early 1870s. The company's first steamers, the **J B WALKER** and **JOHN BRAMALL**, entered service in 1875 and 1876 and were followed by the **DANISH MONARCH**, **SAXON MONARCH** and **NORMAN MONARCH**, which were also employed on the routes to the Far East.

Tureen manufactured for the company, probably by Ashworth Brothers.

In 1879 the decision was taken to enter the passenger and cargo service between London and New York and four larger steamers were ordered to serve this route, namely the **ASSYRIAN MONARCH**, **PERSIAN MONARCH**, **EGYPTIAN MONARCH** and **LYDIAN MONARCH**. Two of these were constructed of iron and two of steel, the **ASSYRIAN MONARCH** being the first steel ship employed on the New York trade. A fifth vessel, the somewhat larger **GRECIAN MONARCH**, came into service in 1882.

Details of the five ships were as follows:-

ASSYRIAN MONARCH (1880/3,317grt) Earle's Shipbuilding & Engineering Co, Hull
EGYPTIAN MONARCH (1880/3,916grt) A McMillan & Son
PERSIAN MONARCH (1880/3,725grt) A McMillan & Son
LYDIAN MONARCH (1881/3,987grt) A McMillan & Son
GRECIAN MONARCH (1882/4,364grt) Earle's Shipbuilding & Engineering Co, Hull

Initially the number of passengers carried was at a satisfactory level. However, by 1886 their numbers had substantially declined and the company was forced into liquidation, the ships being sold in the middle of 1887.

In addition to the passenger steamers, cargo vessels were also employed on the London to New York route.

LYDIAN MONARCH.

SOUTH AMERICAN SAINT LINE

The South American Saint Line operated some of the finest and most modern cargo passenger liners to be built in the United Kingdom on a liner service to the east coast of South America.

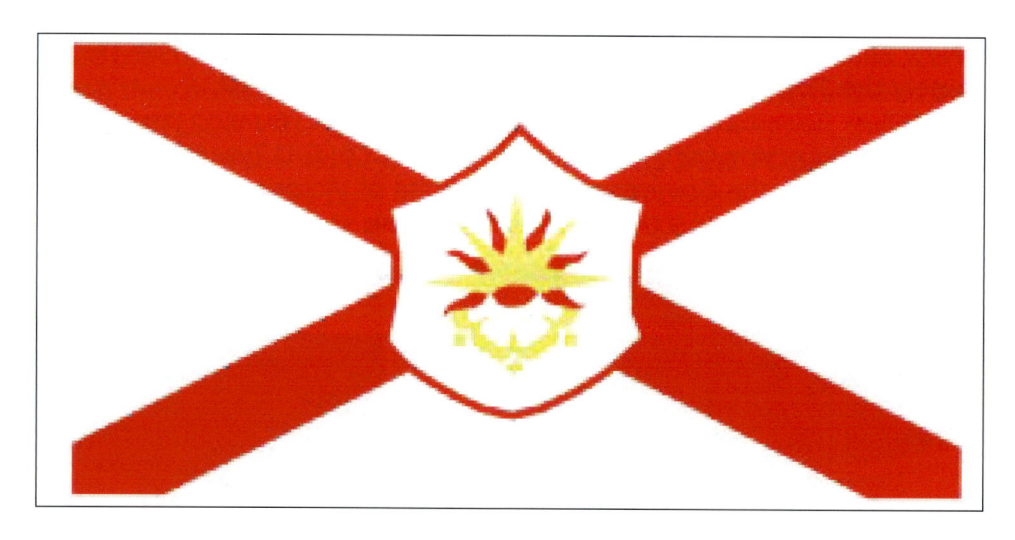

House flag of the South American Saint Line.

In 1926 a Captain George Buchanan Bailey of Newport, formed a single ship company, Monkswood Shipping Co Ltd, to operate a tramp steamer of the same name. The same year the Barry Shipping Company was formed in Cardiff under the management of Richard George Meredith Street. In 1933 Captain Bailey and R G M Street jointly formed B & S Shipping Co Ltd to manage the St Quentin Shipping Co Ltd, owner of a steamer of the same name, and the **MONKSWOOD**.

In 1935 the Government introduced the 'Scrap & Build Scheme' (one ship to be built for every two scrapped). Four ships were built for B & S Shipping under this scheme between 1936 and 1938, these all carrying names beginning with the prefix ST eg **ST. HELENA** (1936/4,313grt).

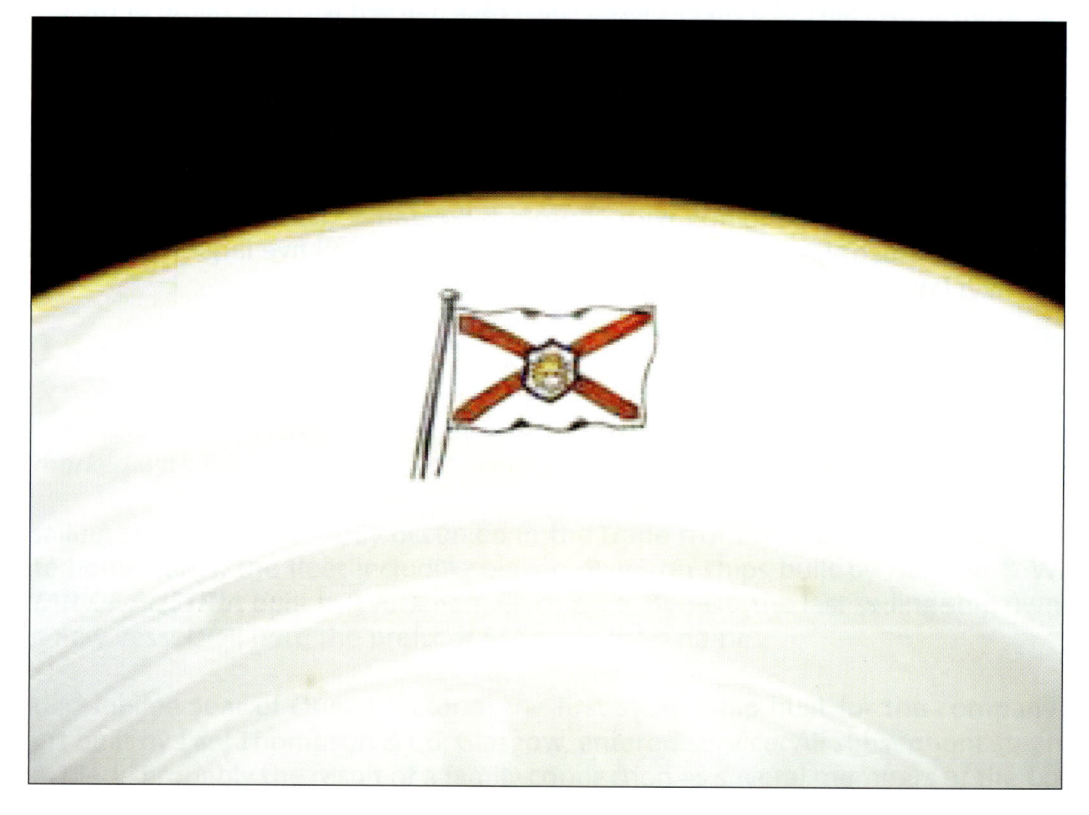

Company dinner plate manufactured by Bristol, England.

Initially these vessels were employed on the coal trade from Wales to Canada returning to Europe with grain but soon a liner service was commenced from Antwerp to Brazil and the River Plate, the vessels calling homewards at London and Hull. The name of the company was changed to The South American Saint Line Ltd in 1939, after Baron Howard de Walden and his son joined the Barry Shipping Company. The vessels were nominally owned by several different companies, the St Quentin Shipping Company Ltd, the Barry Shipping Company and the Triton Steamship Company.

Apart from the cargo liners, B & S Shipping also owned four tramp steamers and entered the war operating nine vessels, a further six vessels were acquired from other owners during the Second World War and the Ministry of War Transport allocated four other ships to their management. In total eleven South American Saint Line ships were lost during World War Two.

ST. ESSYLT (2) built by Joseph L Thompson and Sons Ltd.

By the cessation of hostilities only four ships were managed by B & S Shipping Co, three by the South American Saint Line and one by the Shakespear Shipping Company.

The fleet was gradually re built and in 1948 two of the most advanced and elegant cargo passenger liners were introduced into the fleet, the **ST. ESSYLT** and **ST. THOMAS**, both vessels being of 6,855grt.

By 1956 the fleet consisted of five new motorships engaged in the liner trade and three tramp steamers. During the group's existence, much of its financial backing came from Lord Howard de Walden. With the death of R G M Street in 1961 and the difficulties associated with operating a liner service to South America, the company was closed and the service was sold to Houlder Brothers in 1965.

ULSTER STEAMSHIP COMPANY LTD
G HEYN & SONS LTD, BELFAST

The patriarch of the Heyn family business was Gutavus Heyn, who was born in Danzig in April 1803. In April 1825 he left Danzig for Liverpool and in 1826 he arrived in Belfast where he became a prominent business figure and Consul for Prussia, the first of a number of such appointments. Amongst his many interests he became involved in the ownership and management of a number of small sailing vessels - schooners, brigs, brigantines, barques and snows. These vessels were employed in providing a shipping service from Ireland to the Baltic ports and east coast of the United States of America carrying grain, flour and timber eastwards and emigrants westwards to a new life in the Americas. The vessels also traded to the Mediterranean and Black Sea. Gustavus Heyn died on 4 December 1875 and in the 1880s three sons joined the family business which then became G Heyn & Sons.

The Ulster Steamship Company was registered in Dublin on 25 August 1877 although the offices remained in Belfast. The new company ran services to the east coast of Canada, the Far East, Europe and Baltic Ports. Voyages to New Orleans started in 1896 and the company began carrying a limited number of passengers at about the same time. Mostly the number of passengers was limited to a maximum of twelve on each vessel, this being common for most cargo liner companies. Above this number the status of the vessels changed with additional staff (including a doctor) and equipment being required.

The company houseflag was based on the Bloody Hand of Ulster - the Red Hand - which is associated with many varied stories from ancient Irish history.

Dinner plate showing the company crest - Booths 'Silicon' china.

Dinner plate a later version of the company china.

The first steamer to be owned by the Ulster Steamship Company was the iron vessel **BICKLEY** (631grt) which was being built for Pile & Company in 1877, but acquired the same year by the Ulster company. She was followed in 1879 by a new build, the iron steamship **FAIR HEAD** (1,175grt). This vessel set the naming system for the fleet, the subsequent vessels being named after Irish headlands and the company being commonly known as the "Head Line".

By the outbreak of war in 1914 the number of vessels in the fleet had risen to seventeen with a total gross tonnage of 51,000 tons and these high class vessels were mostly employed on line voyages, a few operating from time to time in the tramping trades. The larger vessels were used on the longer transatlantic trade, the smaller on coastal and continental routes. The company sustained grievous losses during the course of the conflict, eleven vessels being sunk by German submarines.

In 1917 the Irish Shipowners Company Ltd, (Thomas Dixon & Sons, Belfast) known as Lord Line, was taken over, including their two remaining ships, **LORD ANTRIM** (02/4,269grt) and **LORD DOWNSHIRE** (99/4,853grt). This company had run sailings between Belfast, Dublin, Cardiff and Baltimore, Rotterdam to Galveston, and Cardiff to Montreal and Quebec and these ships retained their names. These services continued using the name Lord Line as a subsidiary of the main fleet.

In March 1920 the Ulster Steamship Company was sold to and became a part of, the Amalgamated Industrials Ltd's conglomerate, with serious financial repercussions for the company as shortly afterwards Amalgamated Industrials was placed into receivership. At this time control of the Ulster Steamship Company was taken over by the Westminster Bank with G Heyn & Sons continuing the management of the vessels. Fortunately, the Heyn family was again in a position to take over the assets of the steamship company by 1933.

Following the changes of ownership and the creation of the Republic of Ireland the company was re-registered in Belfast in 1924. At the outbreak of the Second World War in 1939 the Head Line fleet consisted of five ocean going vessels each of approximately 5,000 tons and five smaller short sea traders. As with the earlier war, the company suffered grievous losses due to enemy action and only two vessels remained at the termination of hostilities.

The Ulster Steamship Company fleet was gradually rebuilt after the war and upon the opening of the Great Lakes to foreign deep sea vessels in 1959 the company established a new service. Ulster Steamship Company acquired shares of Donaldson Line together with one vessel - the **SANTONA** - in 1967 and the company traded under the name of Head-Donaldson Line, although each company retained its own identity.

In 1979 the last ship was sold and this marked the end of the Ulster Steamship Company which became dormant although G Heyn & Sons continued to retain many interests in the shipping industry.

TORR HEAD *(1937/5,021grt) built by Harland & Wolff, Belfast - a typical Head Line vessel.*

VILLAGE SS CO LTD
(R A MUDIE & J H MUDIE, DUNDEE)

The Village SS Co Ltd was formed in 1905 to build the steamer **DRUMGEITH** (3,883grt). R A Mudie had bought his first screw steamer in 1866, his ships being named after periods of the day for example **GLOAMIN**.

Large oval platter
– manufacturer unknown showing the company houseflag and the name of the steamer DRUMGEITH.

*The **DRUMGEITH** was sold in 1916 and the Village SS Co Ltd wound up.*

WARREN LINE
(WHITE DIAMOND STEAMSHIP COMPANY)

The Warren Line's foundation lay in the days of the American transatlantic packets running between Boston and Liverpool, the company being established by Enoch Train in 1839. During 1848 an office was opened in Liverpool, George Warren moving from the USA to Liverpool in 1853 to manage it. Eventually he became in full control of the company and his name became that of the line.

During the American War of Independence the ships were transferred to the British flag and the company became a British entity. At this time all vessels consisted of sailing ships, chartered steam vessels first being introduced in 1863 and the first owned steamers entering the fleet from 1875 onwards.

China soup plate marked with the house flag and line name. Supplied by D A S Nesbitt & Co. Manufactured by Cauldon Registration number R.171711 Circa 1891.

The first owned steamer was the **MASSACHUSETTS**, previously the Guion Line vessel **MANHATTAN** of 1866. Frequent sailings were offered, cattle shipments to Liverpool being one of the main cargoes carried. In 1904 a new service was commenced from Galveston to Liverpool, cotton being the main commodity shipped on this route.

In 1898 Warren Line (Liverpool) Ltd was established as the White Diamond Steamship Company Ltd and by 1912 had a somewhat ageing fleet of four ships. George Warren, son of the original George Warren, was looking to retire and the company and its remaining four vessels were purchased by Furness Withy and Company. At this time the company operated services from Liverpool to St John's, Newfoundland, Halifax, Nova Scotia and Boston. During 1913 two of the former Warren Line vessels, **SAGAMORE** and **SACHEM**, were each fitted with accommodation for sixty 2nd class passengers.

In 1934 Furness Withy decided to merge their Liverpool fleets and the Johnston Warren Lines was formed in December 1934. The Warren Line ships were named **NEWFOUNDLAND** and **NOVA SCOTIA**. These names were later given to new passenger cargo vessels under Furness Withy ownership in the 1950s, which were operated to eastern Canada.

*Postcard of the **BEECHMORE**.*

WEST INDIA & PACIFIC STEAMSHIP COMPANY

The West India & Pacific Steam Ship Company was created in Liverpool in 1863 to provide sailings to the West Indies & Central America. The first steamer to operate this service was the 1,880grt **MEXICAN**. She was joined in 1854 by a number of steamers previously owned by Alfred Holt and Company. These had initially been used by Holt's on a service to the West Indies. They had now decided to concentrate on the trade with the Far East (the famous Blue Funnel Line) and within the year the West India & Pacific fleet had risen to a total of fourteen steamers. In 1865 the company obtained a Royal Mail contract to carry mail to Honduras and Mexico, all services operated by the company being included in the contract by 1868.

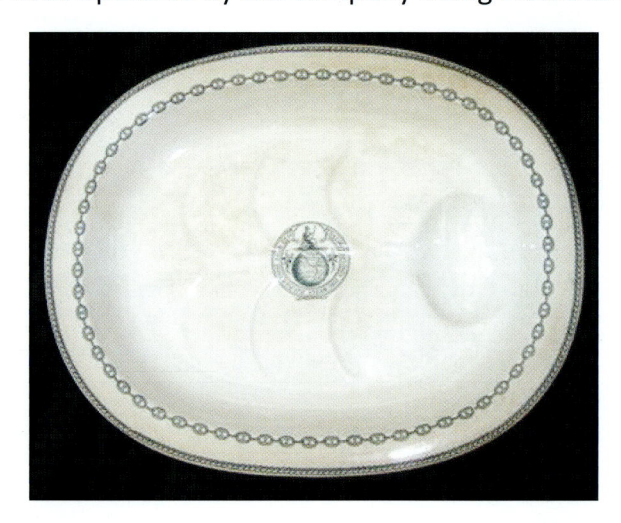

Large well and tree meat plate with the line name in green and a border of anchor chain in green, made by Morley & Harrald circa 1863.

The company continued to prosper and the largest ship to be owned joined the fleet in 1899. She was the **ATLANTIAN** (9,399grt) built by Armstrong, Whitworth & Co at Newcastle.

Largest ship built for the West India & Pacific Co. was the Atlantian, delivered by Armstrong, Whitworth & Co., in 1899. She joined the Leyland fleet in 1900 and was torpedoed and sunk off Eagle Island on June 25, 1918.

In 1899, John Reeves Ellerman, who had taken over the Leyland Line, commenced negotiations with the West India & Pacific Co to merge with the Leyland Line and in December 1899 a new firm Frederick Leyland & Co (1900) Limited was formed to take over the newly merged companies. The nineteen West India & Pacific vessels, totalling some 80,000grt, were transferred to the new line and this marked the end of a successful shipping company that had served the Caribbean and West Indies for 36 years.

CHAPTER THREE

COMMONWEALTH COMPANIES

Adelaide Steamship Company

Anchor Shipping and Foundry Company, Nelson

Australasian Steam Navigation Company

Australasian United Steam Navigation Company Limited (AUSN)

Australian Oriental Line

Burns, Philp & Company

Huddart Parker Limited

Melbourne Steamship Company, Melbourne

Northern Steam Ship Company, Auckland

Pickford & Black, Canada

Quebec Steamship Company

ADELAIDE STEAMSHIP COMPANY LIMITED

The Company was formed in September 1875 in Adelaide, South Australia, by a group of woolgrowers and businessmen, some already having steamship interests in the Spencer Gulf, namely Federal Wharf Co Ltd, Port Adelaide Dredging Company Ltd and Spencer Gulf Shipping Co Ltd. The Adelaide Steamship Company was incorporated on 8 October 1875 to operate services between Adelaide and Melbourne. The first vessels of the new company were the sister ships **SOUTH AUSTRALIAN** and **VICTORIAN**, these vessels being built by D & W Henderson, Glasgow and each was of 716 gross tons.

The company promoters and founding directors included Andrew Tennant, Robert Barr Smith and Thomas Elder of Elder Smith & Co Ltd. Like the founders of many of the Australian coastal shipping companies, most of the directors had been born in Scotland. In July 1876 the company's leading promoters amalgamated their private ship-owning interests to form the Spencer's Gulf Steamship Co Ltd, trading in South Australian coastal waters. The first ship of the new company was the **FLINDERS** which arrived in Australia from England under the command of Captain Hay on 14 March 1875. She was of 521 gross tons and built by J Laing, Sunderland in 1874.

This company and the Adelaide Steamship Company amalgamated in December 1882 and the combined fleet circled the Australian coast from Derby in northern Western Australia to Cairns in northern Queensland. Shipping operations were supported by a large network of agency offices in almost every major Australian port. Over the next decades the company became very successful, albeit with the normal highs and lows associated with a business venture and provided an important coastal service between Western Australia and ports to the East, for both passengers and cargo.

Large meat plate circa 1890 with the impressed mark of Brown Westhead Moore & Co.

A number of notable vessels were brought into service and many of these ships gave valuable service to the British Empire in both World War One and World War Two. Several were requisitioned for Government services as hospital ships or troopships, notably the three sisters **WARILDA**, **WANDILLA** and **WILLOCHRA** in the First World War. These three ships were each of approx 8,000grt and built by William Beardmore & Company on the River Clyde. Other vessels were also taken over for government service.

Attractive china used on the Adelaide Steamship passenger liners.

By the start of World War Two the company owned thirty ships and the company was again forced to surrender nine ships to the Navy, including the passenger liners **MANOORA** and **MANUNDA** which became an Armed Merchant Cruiser and a hospital ship. These ships of some 10,000 tons were also built in Scotland, both by Alexander Stephens & Sons, who built a number of notable ships for the company. **MANUNDA** was in Darwin harbour during the Japanese bombing of the port and was able to bring 260 military and civilian casualties to safety in Fremantle. In all, during the war she carried about 30,000 sick and wounded back to Australia from the Middle East and New Guinea. During the 1940s, a decline in trade necessitated the company to diversify and they began to acquire interests in other companies and projects. Consequently, after the war, the company diversified into towage, shipbuilding, and the shipping of salt, coal and sugar.

*Company postcard of the **MANUNDA**.*

On 20 January 1915 the Adelaide Steamship Company took over the Coast Steamships Limited company, and kept it running as a subsidiary that retained its own identity until 1968.

AUSTRALASIAN UNITED STEAM NAVIGATION COMPANY LIMITED
(AUSN)

From the very beginning and well into the 20th century, the economy and infrastructure of Australia depended on the cargo and passenger services provided by Australian shipping companies, large and small. Without these essential services the country would have been slow to develop and interstate transport slow, if not impossible. Some of these companies were small and short lived, others developed over the years into large and well-known lines, the Australasian United Steam Navigation Company (AUSN) being one of the most important firms.

The company could trace its ancestry back to the Hunter River Steam Navigation Company which was created in 1839 to provide a service from Sydney to the Hunter River, the company being reformed in March 1851 as the Australasian Steam Navigation Company. In 1883 the Queensland Steam Shipping Company came into being with the support of the British India and Queensland Agency, agents for the British India Steam Navigation Company in Queensland. This company was so successful that in 1886 it was strong enough to purchase the fleet and services of the Australasian Steam Navigation Company. In 1887 the two companies were amalgamated to form the Australasian United Steam Navigation Company (AUSN). The funnel colours of the new company were similar to those of the British India Steam Navigation Company, two white bands on a black funnel.

The company houseflag as depicted on the lines later crockery. The flag was devised to incorporate the original white St Andrew's cross of the Queensland Steam Shipping Company and the two blue and red segments of the Australasian Steam Navigation Company.

Over the period of the next 74 years the AUSN operated around the Australian coast, providing essential mail, passenger and general cargo services and during this time owned many notable vessels which gave great service during war and peace.

China dinner plate made by A J Wilkinson & Co, circa 1896.

Two of the best known vessels were passenger ships that had been originally built for the Khedivial Mail Steamship Company as the **FAMAKA** and **FEZARA**. Both were turbine steamers built by Alexander Stephen & Sons in 1922-23 and these were given the names **ORMISTON** and **ORUNGAL** when first chartered by AUSN in 1927, being fully purchased in 1936. Each vessel was of approximately 5,830grt.

*The **ORUNGAL** - This vessel was lost when she ran aground on a reef close to Barwon Heads, near Port Phillip Heads whilst leaving Sydney in heavy rain on 21 November 1940.*

The **ORMISTON** gave invaluable service and during the Second World War she came under Australian Government control, surviving a torpedo hit from a Japanese submarine.

ORMISTON.

After the war ended the vessel continued in service. However, by 1955 she was no longer economical to operate and she was sold to a Greek company that year, the last of the company passenger ships.

The AUSN continued to trade for a few more years, but by 1961 the fleet had shrunk to a mere four cargo ships and the last cargo ship, **CORINDA**, was disposed of in March 1961, the company finally being wound up in 1962.

AUSTRALIAN ORIENTAL LINE

The Australian Oriental Line was formed in 1912 by G S Yuill & Co to take over the services of the John Swire-operated China Navigation Company, in Australia. The first ship to be owned by the new company was the **GUTHRIE**, which was purchased from Burns, Philp & Co. She was followed by the 1886 built clipper bowed steamers **CHANGSHA** and **TAIYUAN**, purchased from the China Navigation Company. However, it transpired that only two vessels were needed for the new company's services and accordingly the **GUTHRIE** was chartered back to Burns Philp & Co and sold in 1913.

The **CHANGSHA** and **TAIYUAN** were transferred to Hong Kong Registry in 1920. However, by now these two vessels were rather outclassed and in 1921 the company attempted to buy the Russian flagged sisters **EMPEROR ALEKANDER III** (1914) and **EMPEROR NICHOLAS I** (1915). The deal fell through because of a mortgage on the ships held by the builders and a lien by French interests.

In May 1924 two new ships were ordered from the Hong Kong & Whampoa Dock Co Ltd, Hong Kong, and the two sisters, both of 4,324grt, entered service as **CHANGTE** and **TAIPING** in 1925 and 1926 respectively. During World War Two both ships were requisitioned by the British Government and put in service for the Royal Navy as victualling/stores issue ships. After the ships were released from Government service they were both refitted and returned to operate on the route from Australia to Hong Kong in 1948 and 1949.

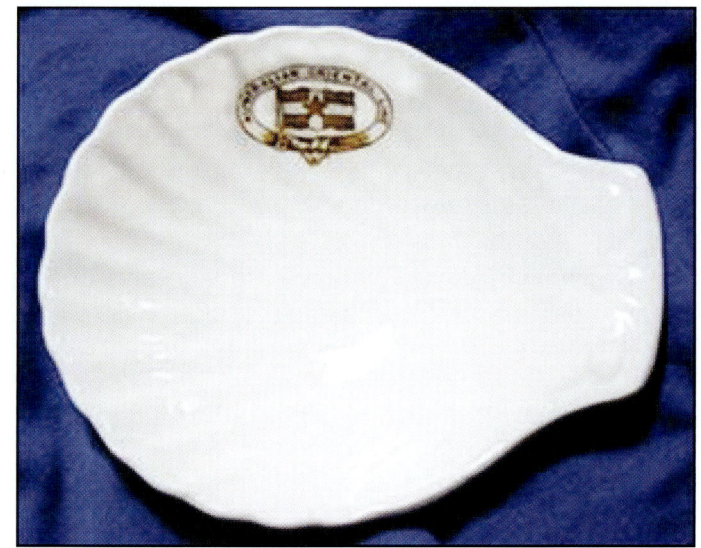

A tea cup manufactured by J H Middleton & Co in their Delphine pattern and a butter dish manufactured by A J Wilkinson.

In 1961 the Australian Oriental Line ceased to exist due to the high costs of acquiring new tonnage or refitting the two ageing existing ships. Service on the route from Melbourne to Hong Kong via Sydney, Brisbane, Townsville, Cairns, Thursday Island and Manila was resumed by ships of the China Navigation Company.

BURNS, PHILP & COMPANY LTD

One of the great trading and shipping companies, Burns, Philp & Co was formed in Sydney in 1876, the result of a partnership between James Burns and Robert Philp. James Burns was born in Scotland and at the age 16 he joined his older brother in 1862 and emigrated to Australia. John Burns founded a retail business in Brisbane and James Burns went to work as a jackaroo in the Australian outback. In 1867 James Burns returned to Brisbane and became a partner in his older brother's business. The discovery of gold in Gympie led the Burns to set up a branch store supplying miners. That store was followed by several others, the business eventually being sold which released capital for other business ventures. James Burns moved to Townsville to establish his own retail and wholesale business in 1872 and two years later, he hired another Scottish immigrant, Robert Philp. Robert Philp continued to run the Townsville store and James Burns moved to Sydney to concentrate on developing a shipping business. Burns, Philp & Company Limited were incorporated in Sydney on 21 April 1883, the two partners being joint Managing Directors. This interest in shipping had been prompted by the need for a reliable service to transport goods for the Townsville store from Sydney and Brisbane, James Burns chartering the small steamer **ISABELLE** to protect his 'supply chain'.

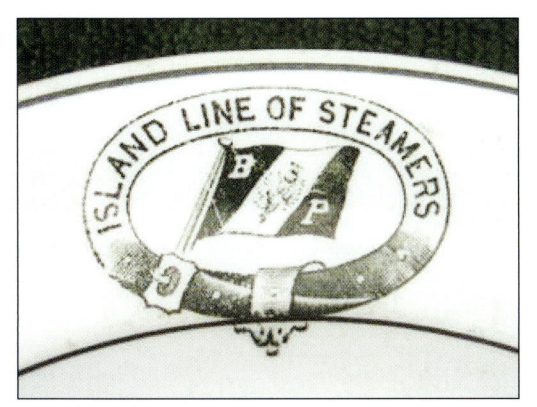

A dessert plate showing the name "Island Line of Steamers" and the company houseflag.

Their business quickly expanded and included a shipping service to Brisbane and Sydney for the inter-colony and overseas trade. In 1885 the company extended its services to the Pacific Islands when it was arranged with the High Commissioner for British New Guinea to provide a monthly service between Thursday Island and Port Moresby. Trading stations were set up at various New Guinea ports and this marked the entry into an area which became central to the Burns, Philp shipping business and to which the company was forever linked. From 1896 a fleet of island traders was based in Sydney and services extended to German New Guinea, Norfolk Island, Lord Howe Island, the New Hebrides, the Gilbert and Ellice and the Marshal Islands and trade developed with the islanders in exchange for copra and island produce. Connections were made at major island ports with the Mail Line vessels to Australia, the company sometimes being described as "The Hudson Bay Company of the South Pacific" acknowledging its great importance to the trade of the Pacific islands.

A fragment of a coffee cup "Burns Philp Line".

By 1914 the Burns, Philp Tourist Department was established, advertising tours on Lord Howe and Norfolk Island. Acquisition of the Port Moresby Hotel occurred in the same year, with the Papua Hotel purchased some years later. Burns, Philp maintained a near monopoly on passenger services to Melanesia until the outbreak of the war in the Pacific. During this period the company had a dominant role in trade in the region distributing general merchandise and collecting copra and provided indispensable links with the Pacific islands.

A postcard of one of the Burns, Philp island steamers sailing from Sydney.

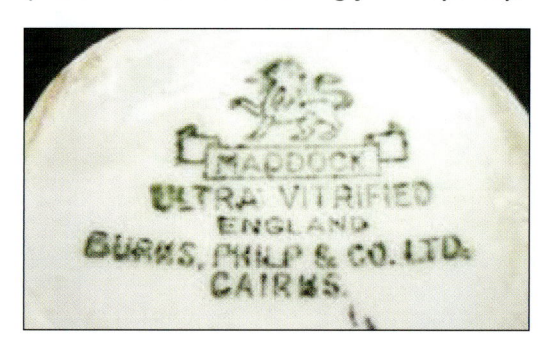

A cream jug supplied by one of the many Burns, Philp stores.

Between the 1880s and early 1970s, Burns, Philp & Co Ltd, operated some thirty-eight main line ships; these were vessels that traded to the mainland of Australia as well as many inter-island sailing and powered vessels. Many of these vessels became household names, not least of which was the **BULOLO**, pictured below. She was launched at the Clydeholm yard of Barclay, Curle & Co Ltd, Glasgow, on 31 May 1938 and had a gross tonnage of 6,267 tons. By the outbreak of the Second World War she had completed eight round voyages from Sydney to the Pacific Islands and on 21 October 1939 she was taken over as an Armed Merchant Cruiser, being commissioned as HMS **BULOLO** on 4 January 1940. Acting in this capacity until 1942, she was then converted to a Landing Ship Headquarters (LSH) and served with distinction at landings in Algiers, Sicily, Anzio and Normandy. She was finally returned to Burns, Philp on 11 June 1948.

The company survived in shipping until 1970, by which time it had disposed of most of its ships and diversified into other industries. Their vessels were easily distinguished by their black funnels with a black-and-white check band and their thistle house flag.

HUDDART PARKER LIMITED

Both Huddart and Parker had emigrated from England to Australia. Captain Peter Huddart had been master of a vessel named **ABERFOYLE** and had settled in Geelong where he opened a business to bring coal from Newcastle in small sailing vessels, later being joined by his nephew. James T J Parker had arrived in Geelong in 1853 where he opened a business as merchant and importer. In 1876 James Huddart & Thomas James Parker, together with a John Trail and Captain Webb formed the business of Huddart Parker & Company. Prior to this, Parker had been involved with a vessel, the **DESPATCH**, operating a successful Geelong-Melbourne cargo and excursion service.

The first coastal passenger vessel built for the company was the iron screw steamer **NEMESIS** built by Turnbull & Son of Whitby in 1880. She served together with two other similar steamers on the route between Newcastle, Geelong and Melbourne carrying both passengers and cargo, including coal. These vessels were also a success and two larger steamers were ordered in 1882, the **BURRUMBEET** and **CORANGAMITE** both built by Swan, Hunter at Wallsend in 1885 and measuring 2,420grt. With these two superior vessels a service was begun between Melbourne and Sydney.

Large water jug and coffee cup and saucer showing the company logo.

Between 1892 and 1898 the company cooperated with the Union Steam Ship Co in a joint transpacific service between Sydney and Vancouver, at first under the name of the New Zealand & Australia Steamship Company and later the Canadian Australian Royal Mail Line.

In 1889 the company was approached by the Tasmanian Woolgrowers Agency Company to provide a vessel for a shipment of valuable sheep from Launceston, Tasmania, to Melbourne and this was the start of a regular service to Tasmania. By this time the fleet comprised a number of steam and sailing vessels and in 1893 the company commenced regular sailings to New Zealand with the passenger steamer **TASMANIA** (1892/2,252grt) and the **ANGLIAN** (1894/2,159grt), a former Union Steamship Company ship previously used on the mail service between Southampton and South Africa. These ships were joined on the route by the new steamers **WESTRALIA** (I) (1897/2,884grt) and **ZEALANDIA** (I) (1899/2,771grt). On 1 January 1912 the company became public and renamed Huddart Parker Limited.

Beautiful plate
(most of the company china was supplied by C Mc D Mann & Company, Hanley, England).

As was the case with many British & Commonwealth passenger shipping lines, several Huddart Parker ships were requisitioned by the British Admiralty for war service as troopships or hospital ships in both world wars. Amongst other vessels in the First World War, the passenger ship **ULIMAROA** (1908/5,777grt) became a troopship in 1916 and the **NAIRANA** (1921/3,042grt) was taken over by the Admiralty, whilst she was under construction and completed as a seaplane carrier in 1917. The **ZEALANDIA** (II) (1910/6,660grt) was requisitioned as a troop carrier in 1918 and was to serve again in World War Two. The **WIMMERA** (1904/3,022grt) was not requisitioned but she was sunk by a German mine off the north of New Zealand.

The company had provided a regular service to Tasmania since 1889 and initially had faced competition from both the Union Steam Ship Company of New Zealand and the Tasmanian Steam Navigation Company on the routes. In 1891 the Union Steam Ship Company took over the Tasmanian Steam Navigation Company and in December 1921 Huddart Parker and the Union Steam Ship Company joined forces and formed a new company, Tasmanian Steamers Proprietary Limited.

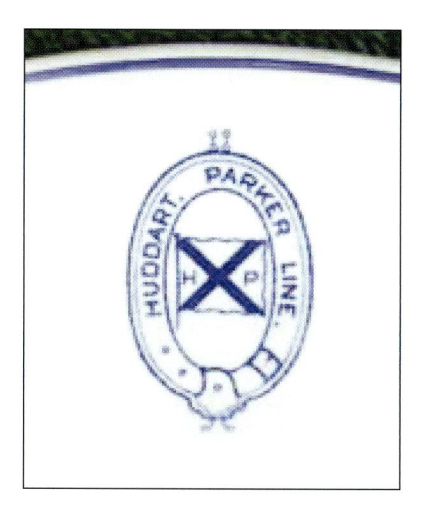

Variations of the company crests on the company china.

In 1939 Huddart Parker owned nine vessels and three of the passenger liners again provided valuable services to the war effort in the Second World War, three liners being requisitioned. These were the **ZEALANDIA** (II) (troopship), **WESTRALIA** (1929/8,108grt) (armed merchant cruiser) and **WANGANELLA** (1929/9.576grt) (hospital ship). The latter had originally been built for the British & African Steam Navigation Company - Elder Dempster as **ACHIMOTA**. This company sold her to Huddart Parker just before her building had been completed. She was a highly successful vessel and remained in service until the demise of Huddart Parker when it was taken over by Bitumen and Oil Refineries Australia Limited in October 1961.

Supplier and manufacturer details.

*Company postcard of the **WANGANELLA**.*

NORTHERN STEAM SHIP COMPANY, AUCKLAND

Company house flag.

As with the Anchor Shipping & Foundry Company, coastal shipping services were urgently required to serve Auckland and the many small ports and rivers in the surrounding area. Several local ship owners, including Captain Alexander McGregor banded together - their first steamship being the **ROWENA** that was built in Auckland in 1872. This ship together with an additional six vessels represented the first fleet of the Northern Steamship Co Ltd. This fleet was operated, following negotiations with the Union Steam Ship Company in 1881 which gave the Auckland parties the right to serve the northern routes. The Northern Steamship Company was formed in 1881 to serve the ports on the east coast of the Auckland province, from East Cape to Parengarenga, and to the Hokianga in the north. On the west coast they served the ports of Onehunga, Waitara, New Plymouth and Wanganui.

Initially the company prospered, but as the result of various mishaps and accidents, a depression and the falling off in the use of timber needed for building, trading conditions in the last years of the 1800s were difficult. Due to these factors, consideration was given to winding up the line. However, Captain McGregor was replaced by a new manager in 1888, Charles Ranson, an English accountant, and by the end of the 1890s the Northern Steamship Company had become one of the main New Zealand coastal companies.

Tea cup and saucer showing the crest of the Northern Steamship Co Ltd.

Apart from cargo, the ships carried a number of holiday makers on short cruises, the first of these occurring in 1903 and they were very popular with the increasingly affluent population. By the 1920s the New Zealand road and rail links had improved and this coupled with a depression towards the end of the decade presented the company with difficult trading conditions and passenger services were terminated in 1929. By the 1930s the company had to lay up eighteen ships although trading conditions had improved by 1937 and the company resumed purchasing new vessels, replacing a number of the steamers with motor vessels. The Northern Steamship Company had a busy period during the Second World War and two vessels were used by the navy as examination vessels, two were sold to the United States and other ships used to carry cargoes for military purposes.

Following the end of hostilities the decision was taken to concentrate on the interisland trade between the North and South Islands, which required larger vessels than those normally operated by the line. Containerisation was introduced in the 1960s Some bulk cargoes of grain were profitable for a time and the company also entered the roll on/roll off trade in 1968 when they acquired the **SEAWAY PRINCESS**. Unfortunately this vessel proved to be unsuccessful, largely due to competition from the Union Steam Ship Company.

In 1974 the company sold its remaining three vessels and ceased operating as ship owners.